Printing 1.

This was produced by Bird Rock Multimedia Corporation 4901 Morena Blvd. Suite #600 San Diego, CA 92117 passmorelab.com

Typesetting by PassmoreLab. Printed and bound in San Diego, CA

ISBN 0-9748822-0-8

Lost and Forgotten: A Visual Documentary of Bethlehem Steel

Photography by: Greg Passmore

CREDITS

We would like to thank:

Advisors: Mr. Donchez and Mr. Martin

Forward: James B. Thomas, PhD

Research: Corinne Kavounas

Layout: Kimberly Lostroscio

Editing: Jennifer Hilbert

Without your help, this book
would not have been possible.

FORWARD

From the 19th Century's Gilded Age through the great industrial machinery of World War II, steel was synonymous with United States industry, strength and power. The images of fiery cauldrons of molten metal, powerful men sweating and hoisting, shoveling and pouring, working in huge factories, all became part of the American consciousness. From rolling rails to carrier keels, no other industry seemed to define and idealize American labor and business in the way steel did. Bethlehem Steel, along with its principal competitor, U.S. Steel, led the way in defining American industry.

Bethlehem Steel (originally Bethlehem Rolling Mills and Iron Company) was created shortly before the Civil War and placed in the ore-rich Lehigh Valley. While the war disrupted initial establishment of the facilities, it also helped generate early contracts and quick success of the company. Under the leadership of John Fritz, and eventually Chairman of the Board Charles M. Schwab, Bethlehem grew rapidly, expanding into a variety of iron markets and became one of the foremost businesses in America. By the 1880s and the introduction of the Bessemer process, Bethlehem took off and became a principal steel manufacturer in the world, and soon entered the world of armaments and battleships. (The Bessemer process is the means by which pig iron is converted to steel by forcing air through the molten iron, oxidizing and separating the impurities and thus creating the stronger, less brittle steel. The operation of the giant converters is generally the image most people have of the steel industry).

By the turn of the 20th Century, Bethlehem was producing steel and cast iron for everything from a sewage disposal plant for Detroit to turbines for Niagara Falls and steel for America's Cup yachts. Millions of tons of iron and steel poured out of the foundries and furnaces of Bethlehem steel.

By World War II, 30,000 men and women were employed at Bethlehem, and America's war effort was based as much on massive industrial production as it was on the arms of the combatants. In this modern, industrial, steel-based war, Bethlehem steel shone in its finest hour. Massive amounts of steel for aircraft engines, ship's armor and big guns all poured out of Bethlehem. If credit for winning World War II were doled out among the participants in the country, the steel workers would be honored shoulder to shoulder with the soldiers, sailors, airmen and Marines. The years following World War II began the steady decline of not just Bethlehem, and not just the steel industry, but much of America's industrial power. Foreign competition, tariffs and a changing world saw the displacement of much United States industry. While still vitally important to the world market, and with a quality still unsurpassed, American-made steel has been reduced to a fraction of the dominance in production it once held.

The last cast for Bethlehem Steel took place in November, 1995. Since then, the plants have been on the inevitable deterioration into rust and memories. The photographs in this collection are a haunting reminder of what once was, and of the direction so much of America's great industry went in the decades after World War II. While looking at these pictures, see them not just as worn out machines but see them for what they once were: a vital, living part of America's greatest moments of industrial might and what American labor could accomplish in its heyday.

James B. Thomas, PhD
HCCS-Northwest College
Houston, Texas

Walkway on Blast Furnace

Incline plane also known as a skip hoist; behind the stairway you can see the skip tub which is actually used to transport the raw materials (coke, arnore and limestone) from the stockhouse up to the top of the blast furnace; there are 5 blast furnaces each having their own skip hoist.

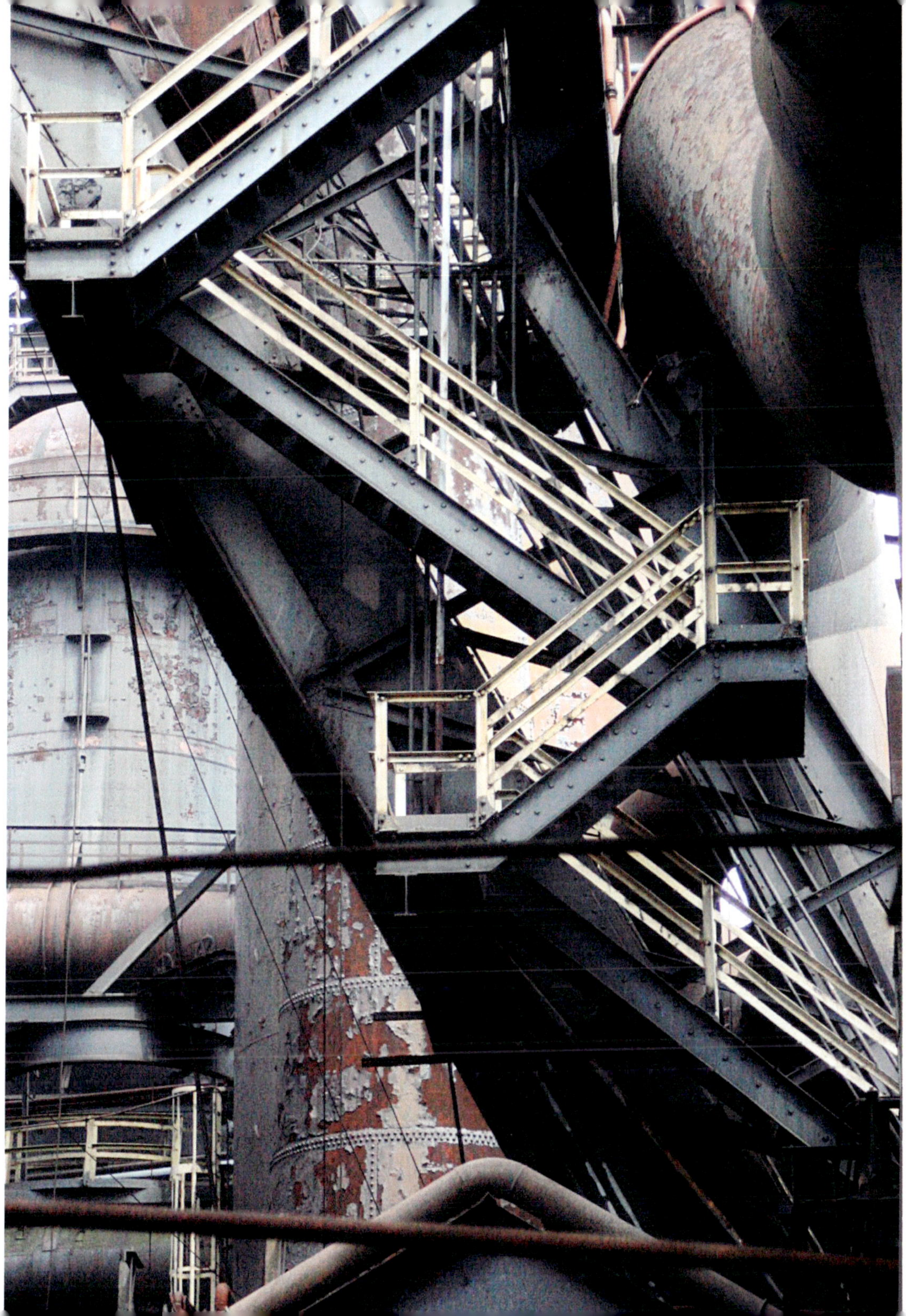

Blowing Engines

Partially taken apart blowing engine in the eastern end of the Lehigh Power House. The power house houses the blowing engines which fuel the blast furnaces; the southern end of the blowing engine is an internal combustion engine which runs on blast furnace gas and converts gas into mechanical energy; the gas engine drives a reciprocating compressor at the other end; the fly wheel is a 100 ton fly wheel (17 total in power house) which makes the wind 30 lb. per square inch of air which then goes into the tweers of the blast furnace; this is a vital support facility for the blast furnace operation for blowing the wind into the tweers of the blast furnace; what is unique about it is that there are these large internal combustion engines for an 8-cylinder engine to drive the 100 ton fly wheel which in turn drives a reciprocating air compressor to fuel any one of the five blast furnaces.

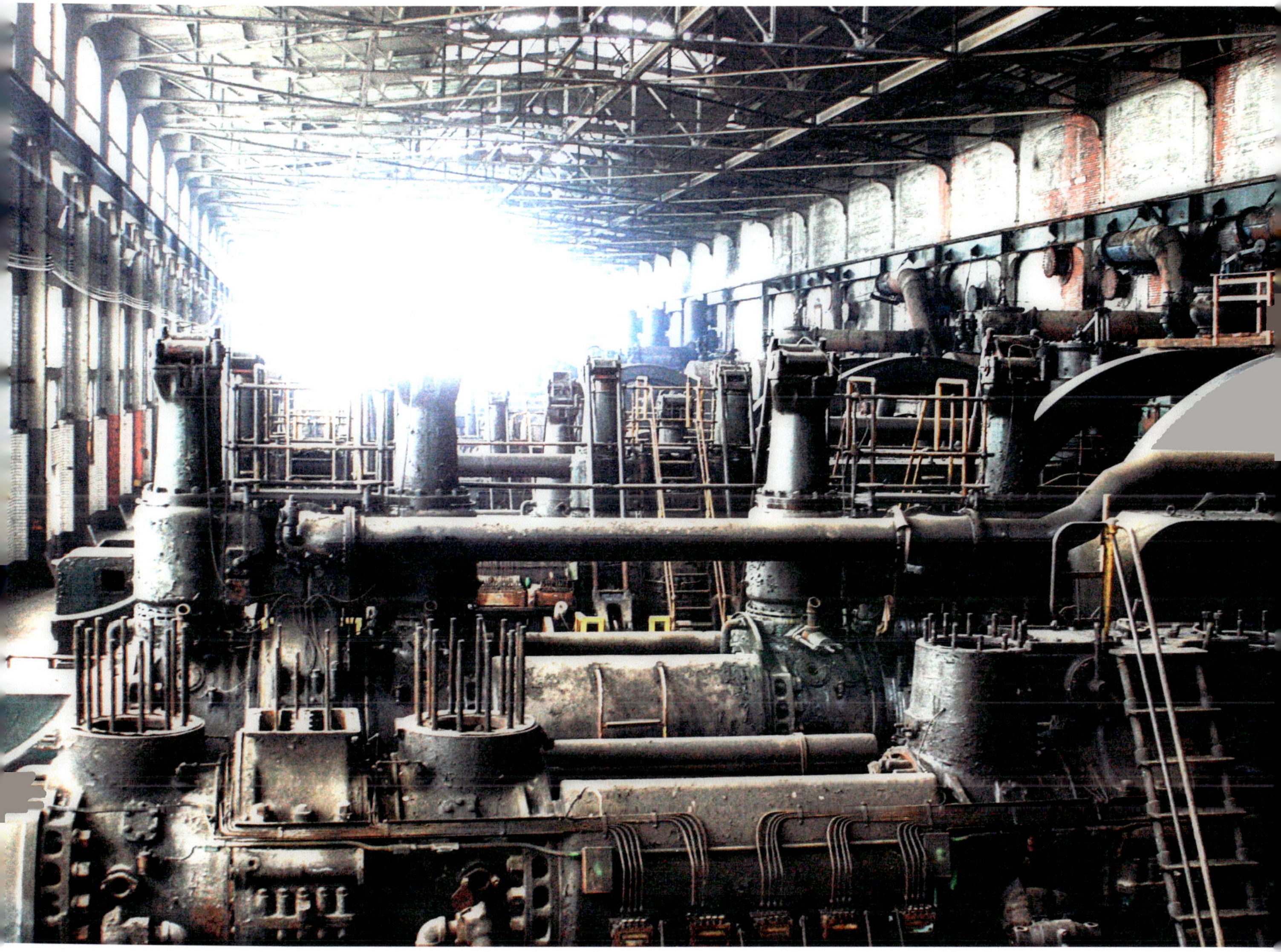

Interior of Number 2 Shop

Large columns inside No. 2 machine shop. There are two types of columns shown in the picture; one of a wide flange column put together in the later part of the 1800's and then a picture of a cast iron column (flares out at the bottom); after 1906 future columns installed were made on the gray mill at the Bethlehem Steel and are not riveted or flared.

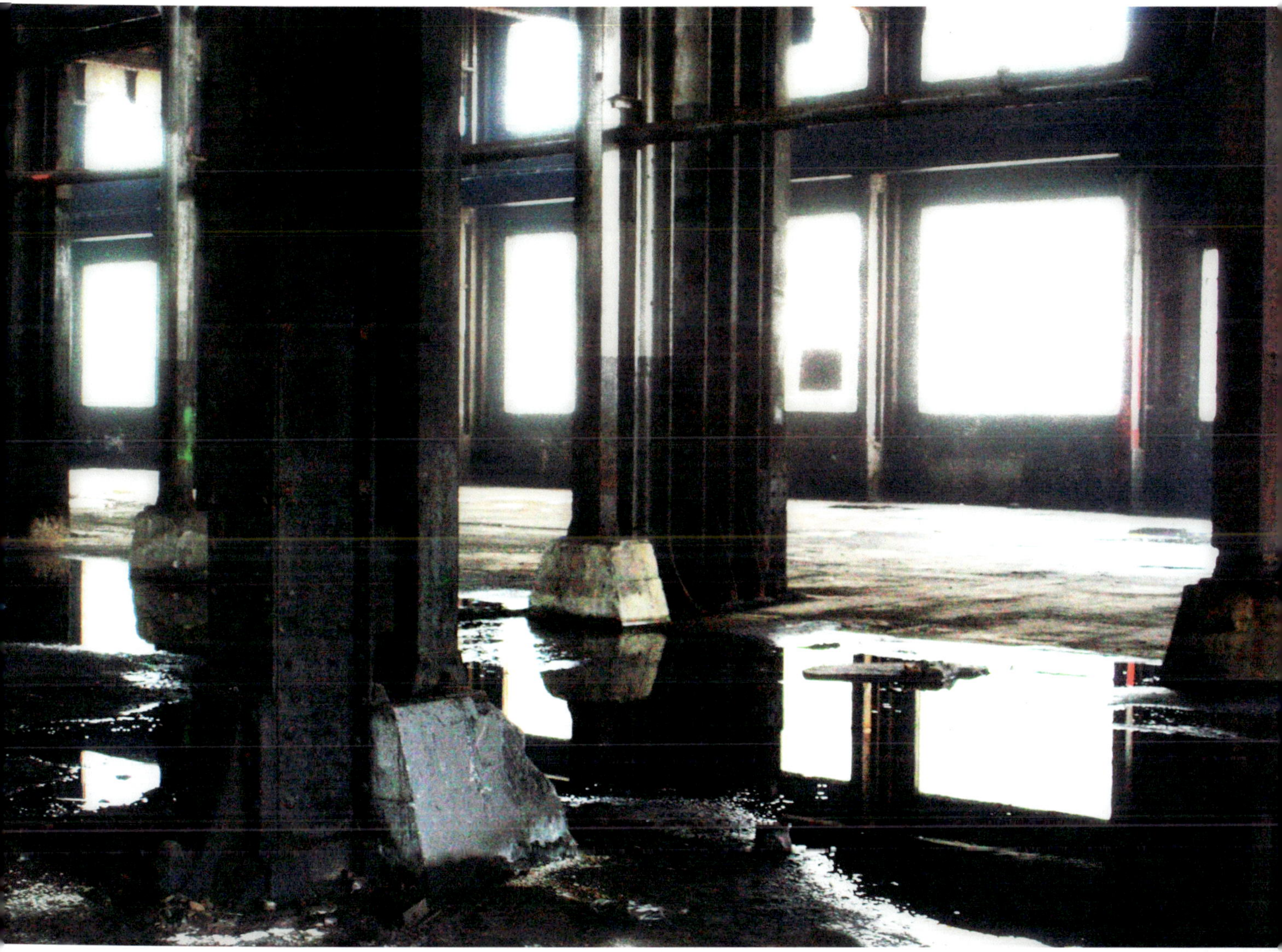

Blowing Engine

Disassembled blowing engine inside Lehigh Power House.

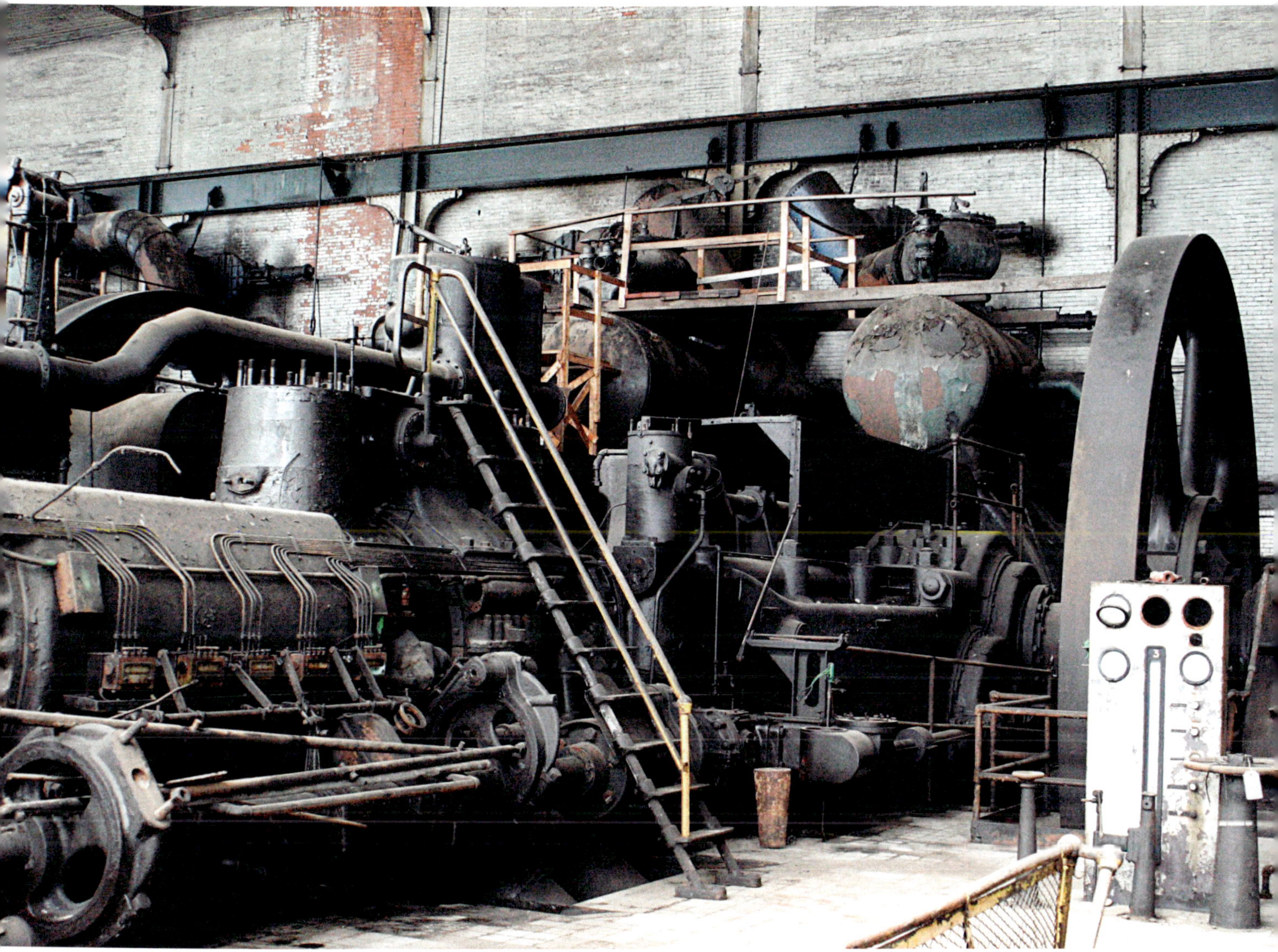

Detail on Blowing Engine

Closeup of lubricator on blowing engine; squirts grease on various bearings on the blowing engine.

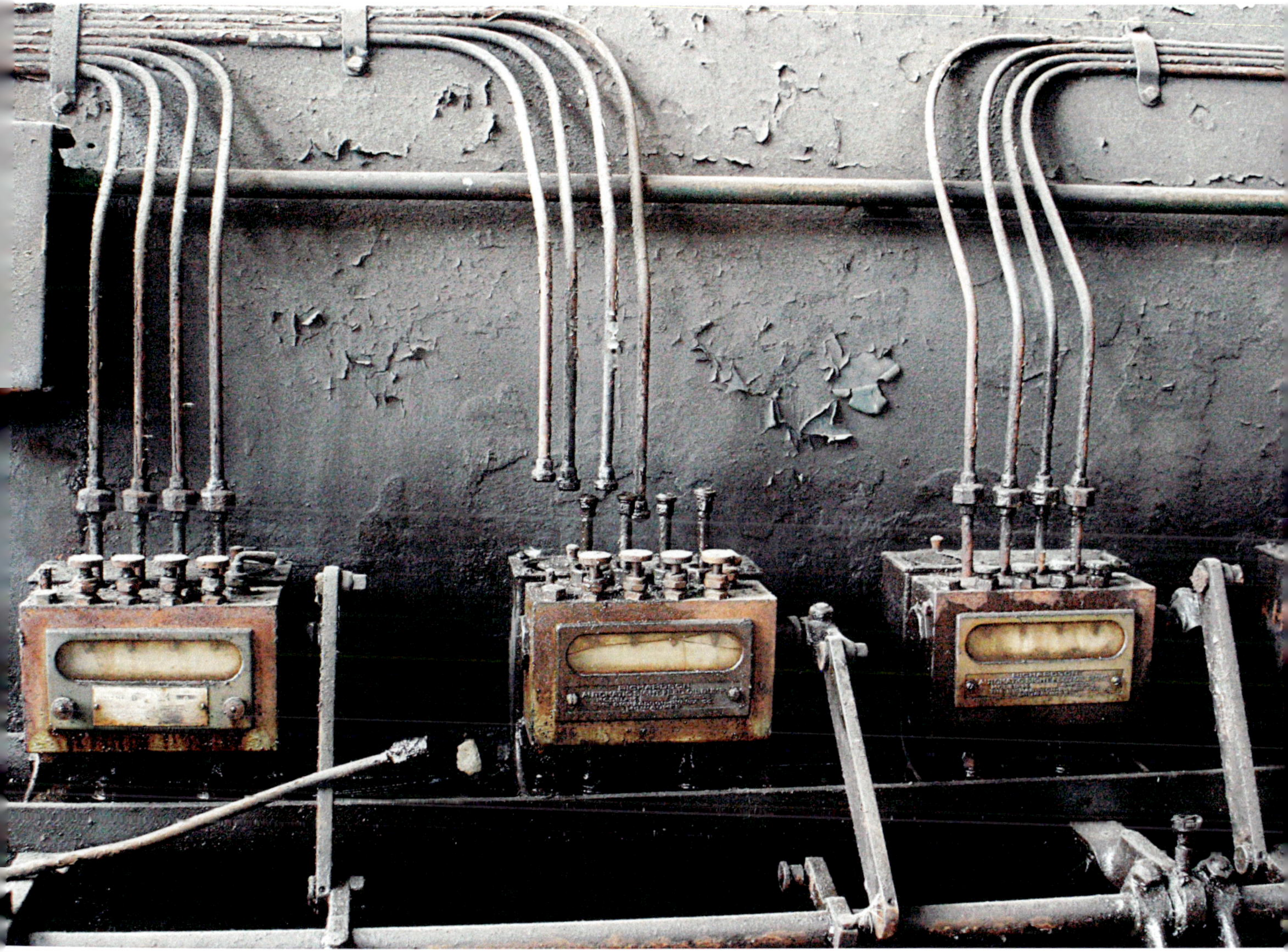

Top Picture:

Control gauge on control panel of blowing engine.

Bottom Picture:

Dirty window

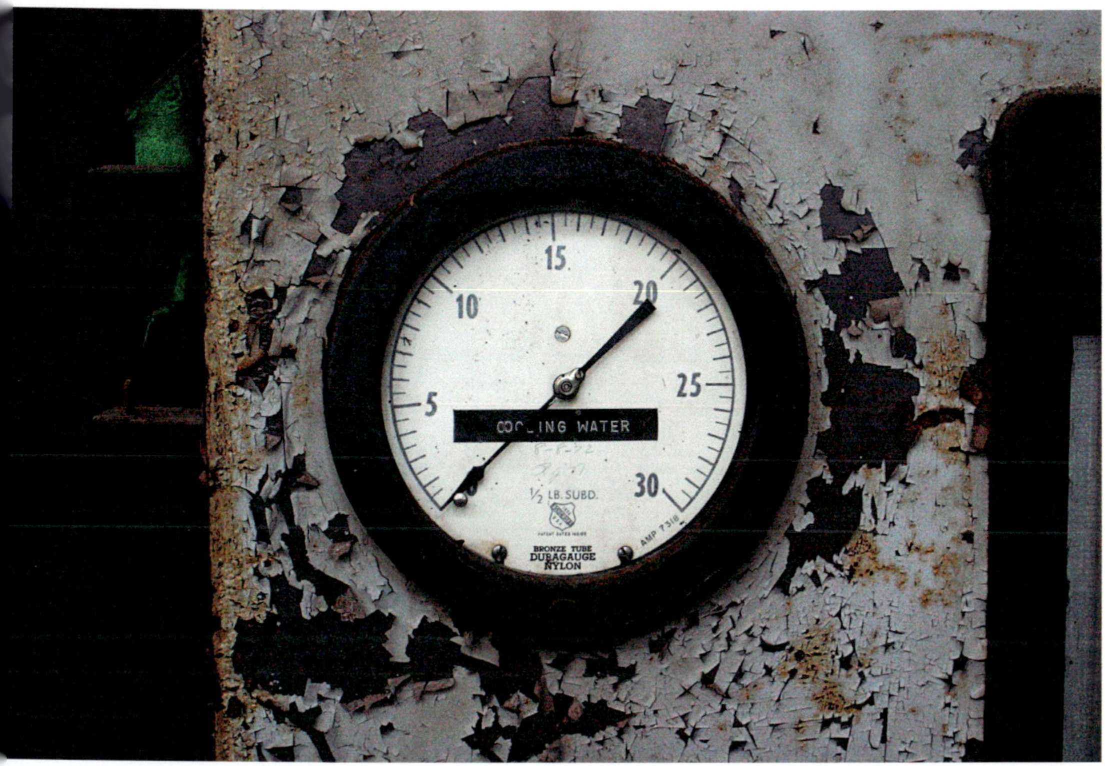
15
10
20
5
25
30
½ LB. SUBD.
BRONZE TUBE
DURAGAUGE
NYLON
AMP 7318

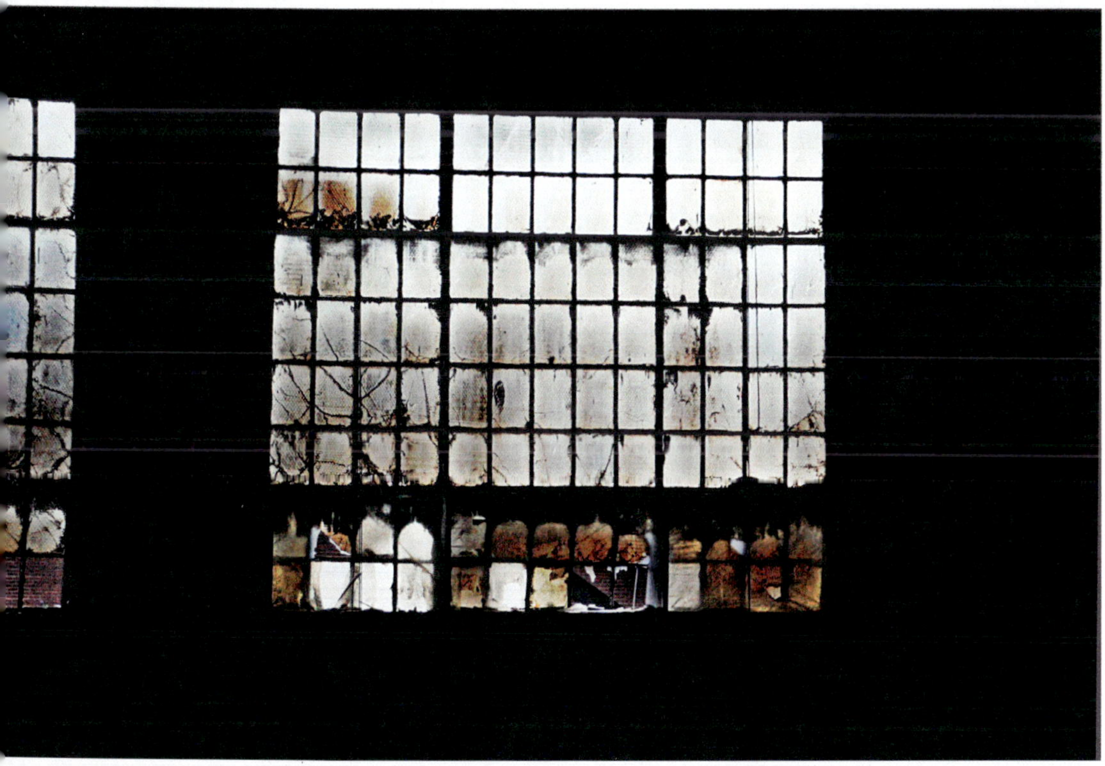

Entrance to Interior of Blowing Engine House

Entrance to Lehigh Power House; south wall showing the unique detail of building made of ceramic brick with quarry tile floor.

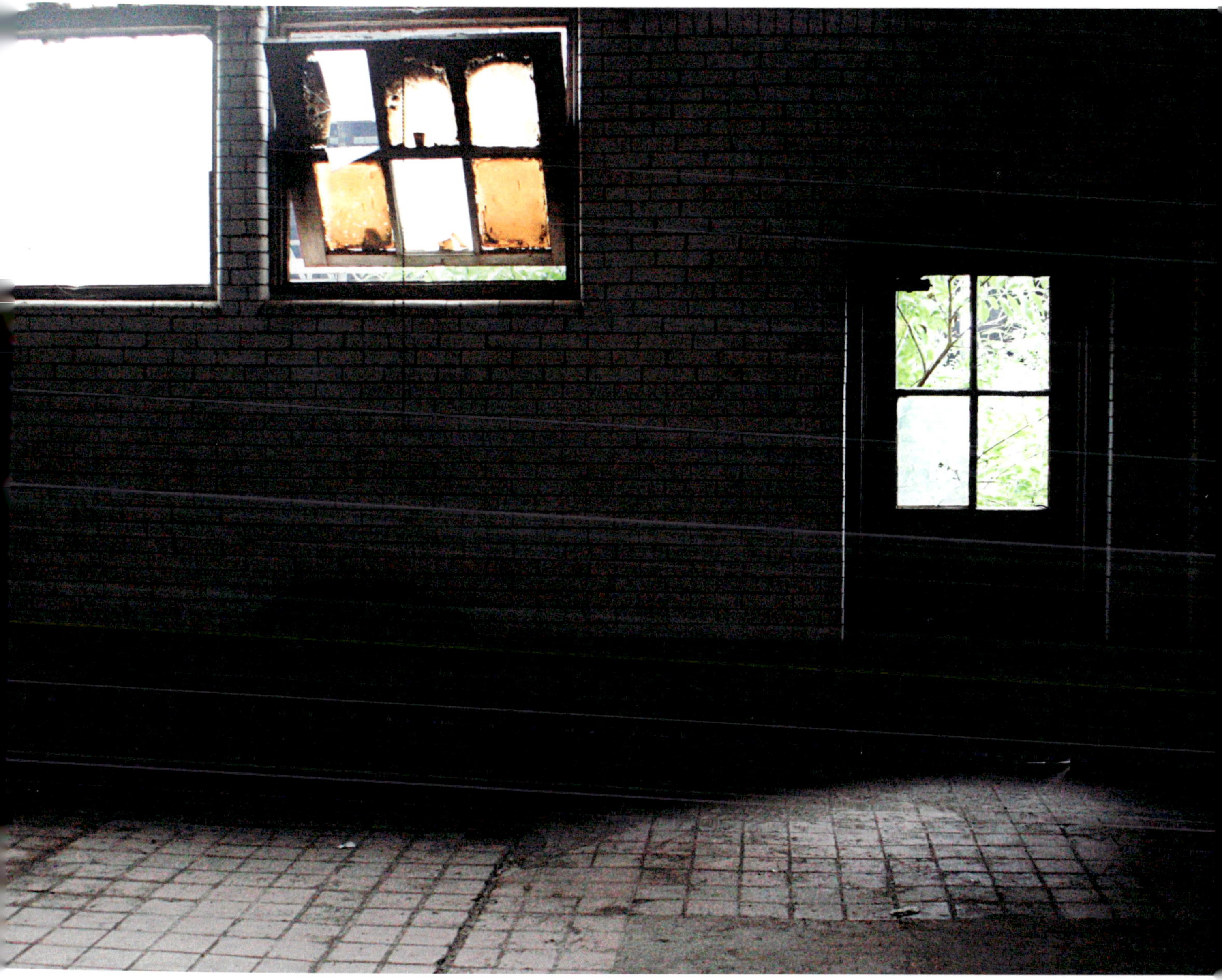

Welfare Room

Showing various types of baskets used to pulley workers' belongings to ceiling and then secure them with a lock; near Number 2 machine shop.

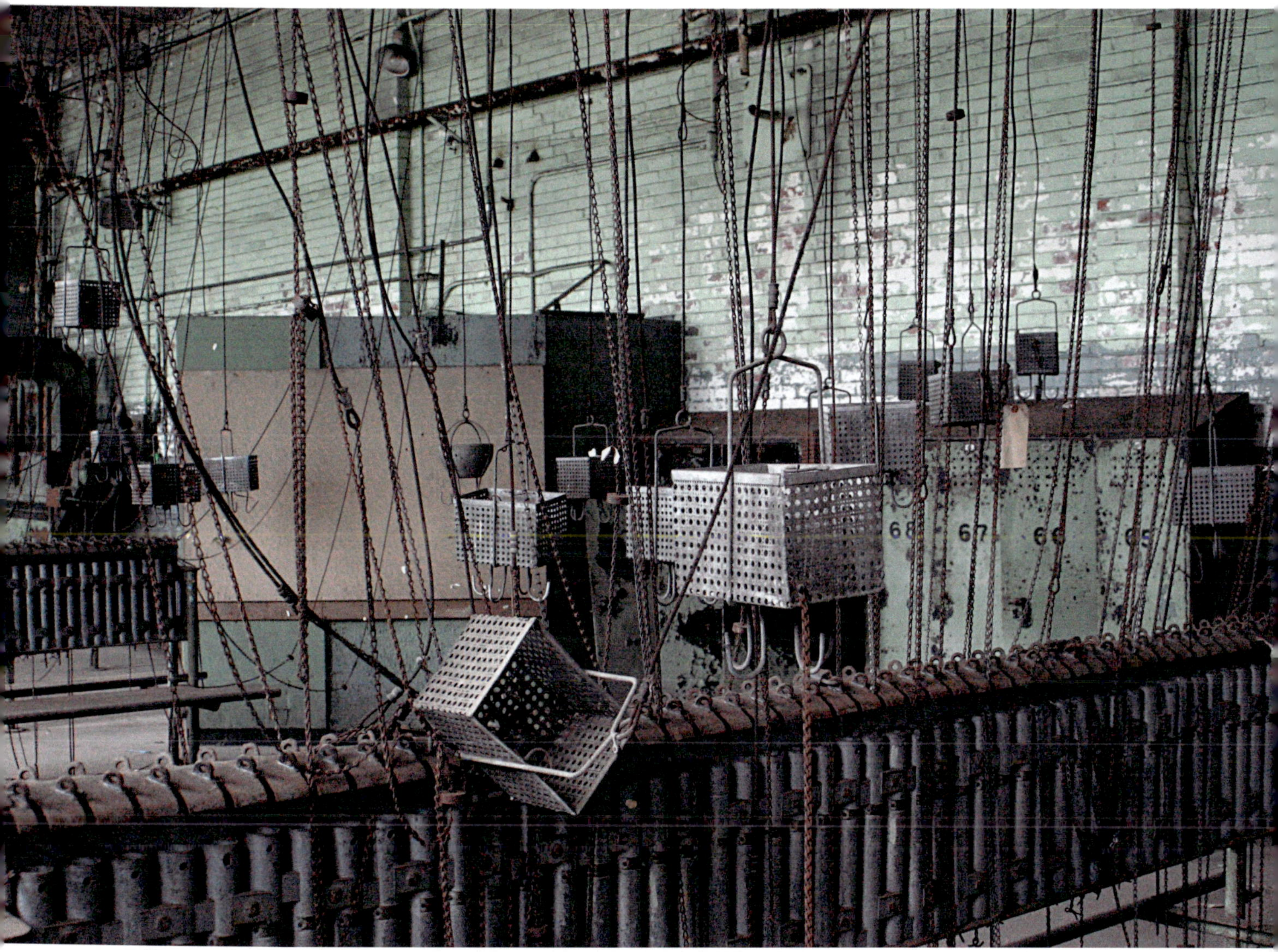
68
67

Number 2 Machine Shop

Inside of Number 2 machine shop looking east; showing higher bay.

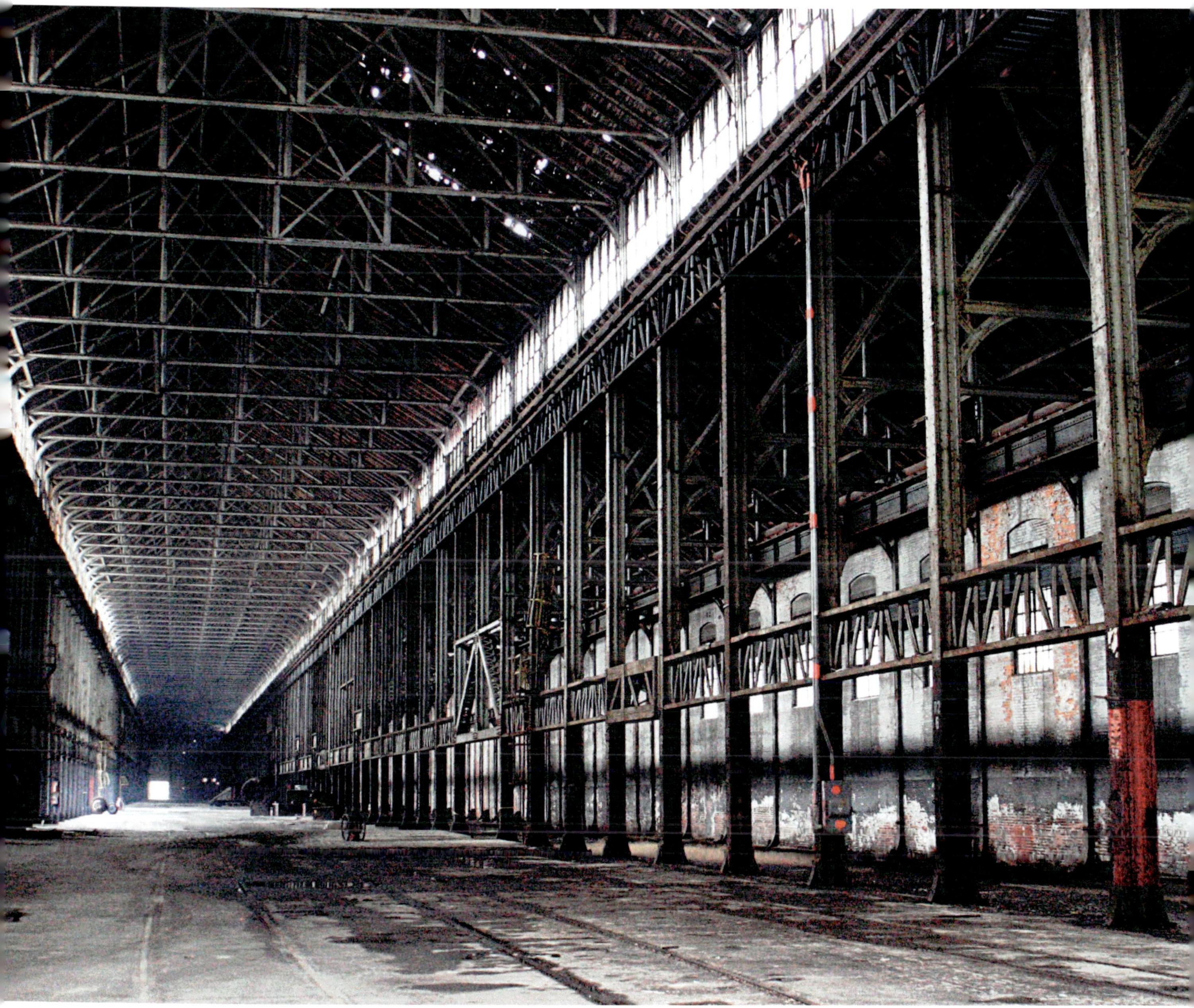

Blast Furnace Skip Hoist

Pictures of skip hoist; cat walks and stove going to the top of the blast furnace; each blast furnace has 4 stoves which preheat the wind before it goes into the bottom of the blast furnace.

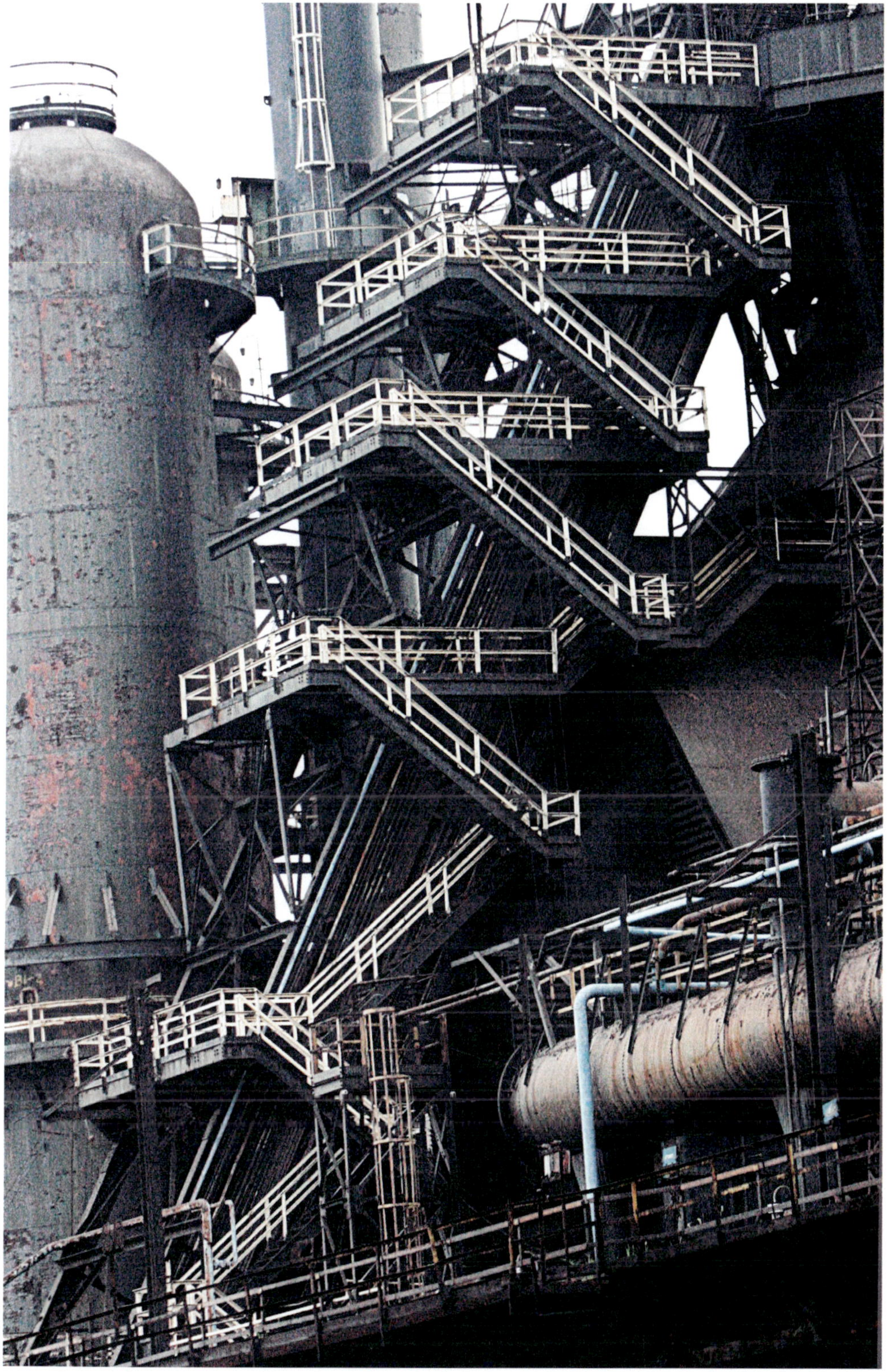

Blast Furnace

Blast furnace gas scrubber tower which cleans the gas before going into the blast furnace.

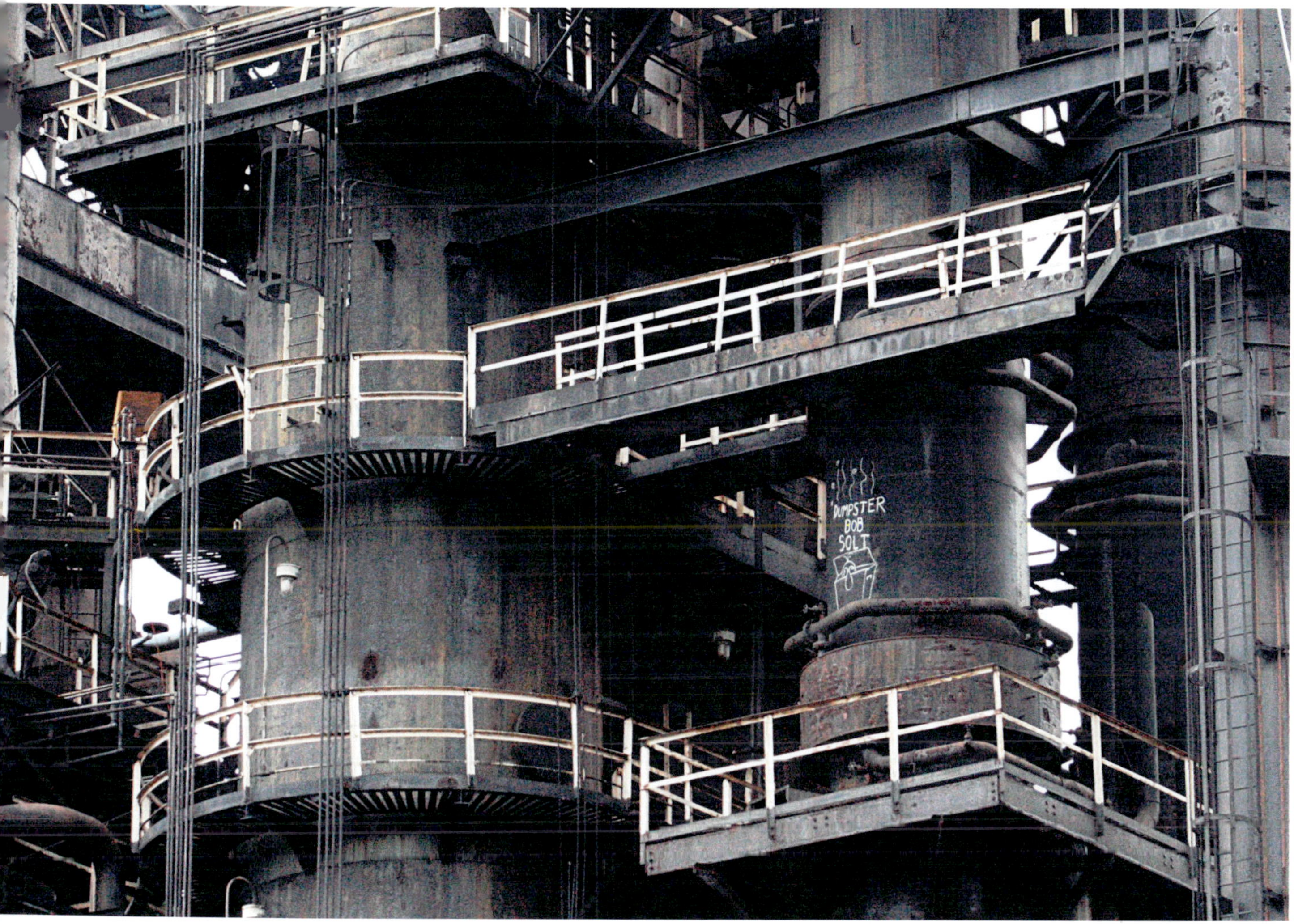
DUMPSTER
BOB
SOLT

Window

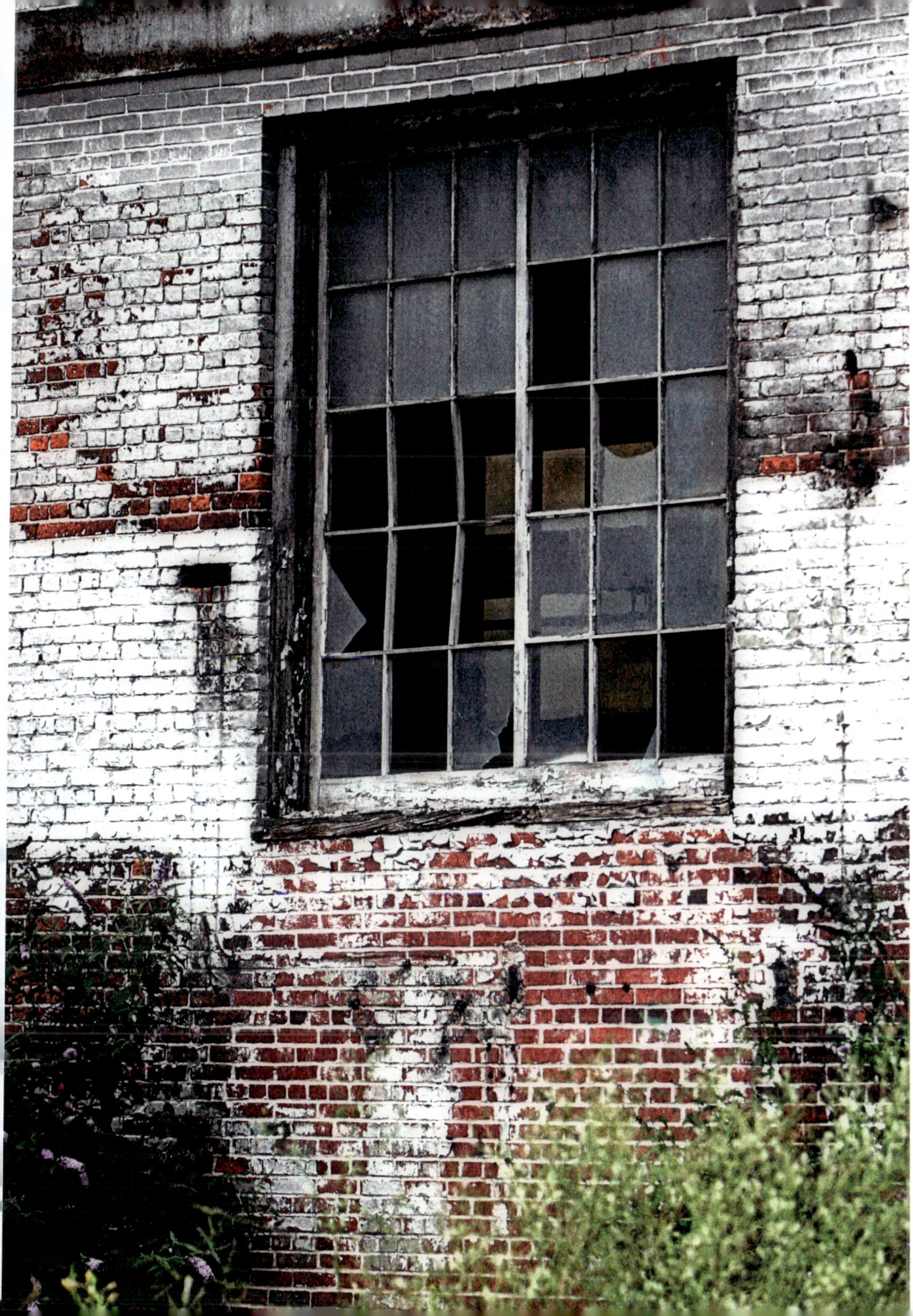

1863 Stock House

Building on the right is Central Tool; building on the left is Central Tool Annex; stone building straight ahead is the original stock house for one of the original blast furnaces; two tall circular images are the stoves for letter A blast furnace; stoves preheat the wind before it goes into the bottom of the blast furnace; the letter A furnace was built and replaced in the late 1800's.

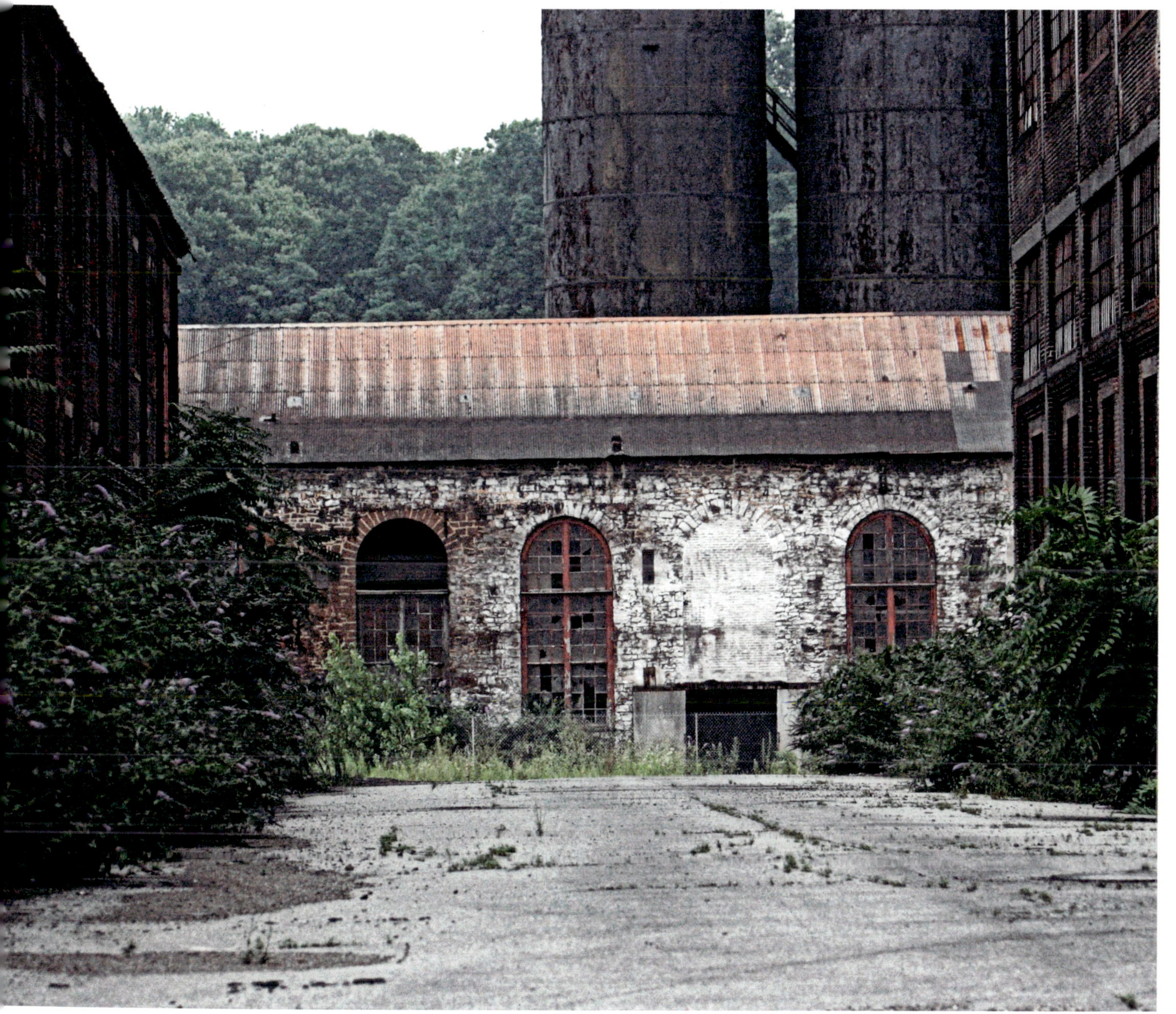

Central Tool

Closeup of Central Tool; at one time was a one story building and then added onto; in 1915 it became a three story building.

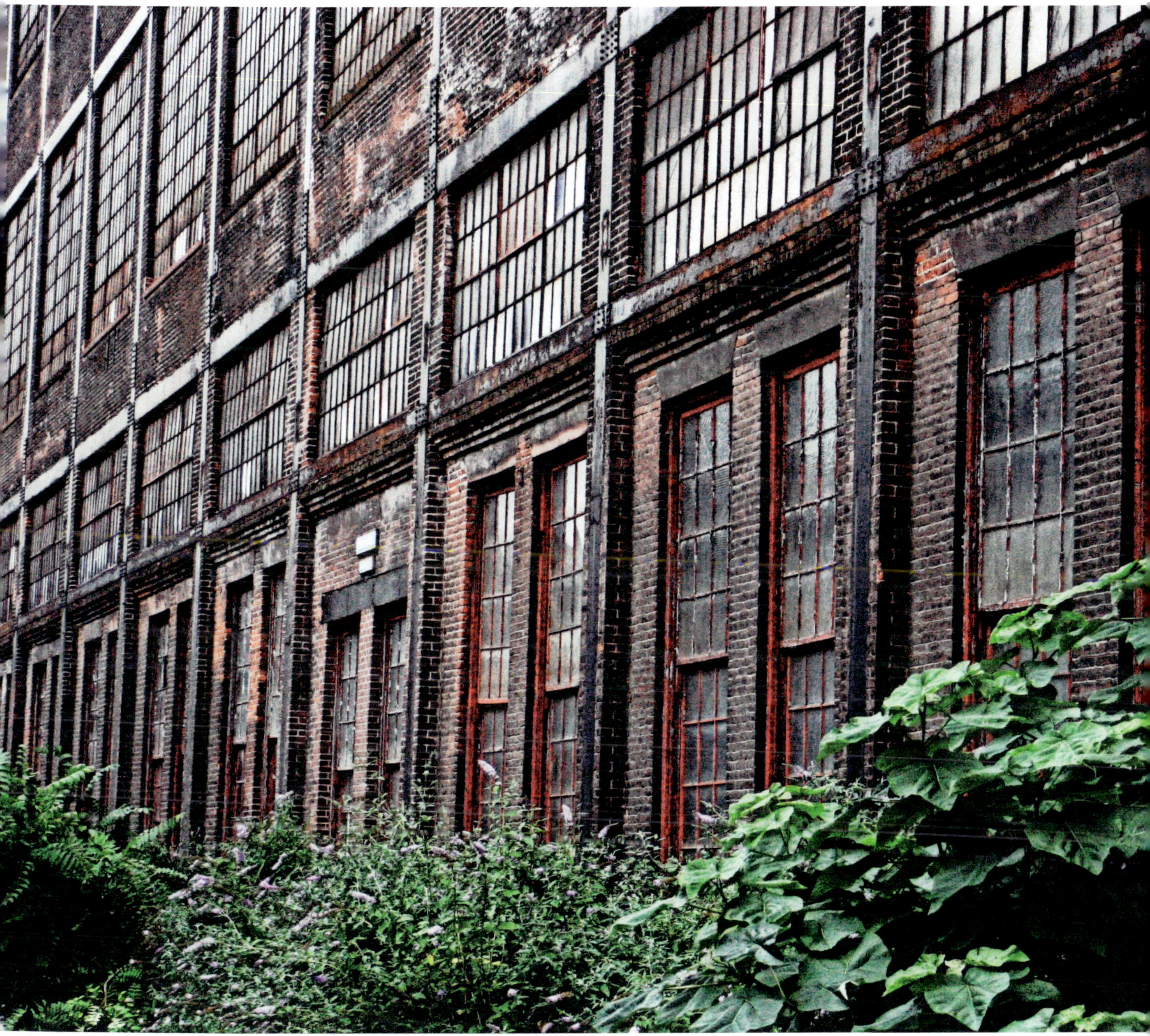

Building 199; entrance to Blowing Engine Room (due to Notice sign)

Building 199; heat treatment building which has a high crane in it; vertical treatment building; No. 3 Treatment.

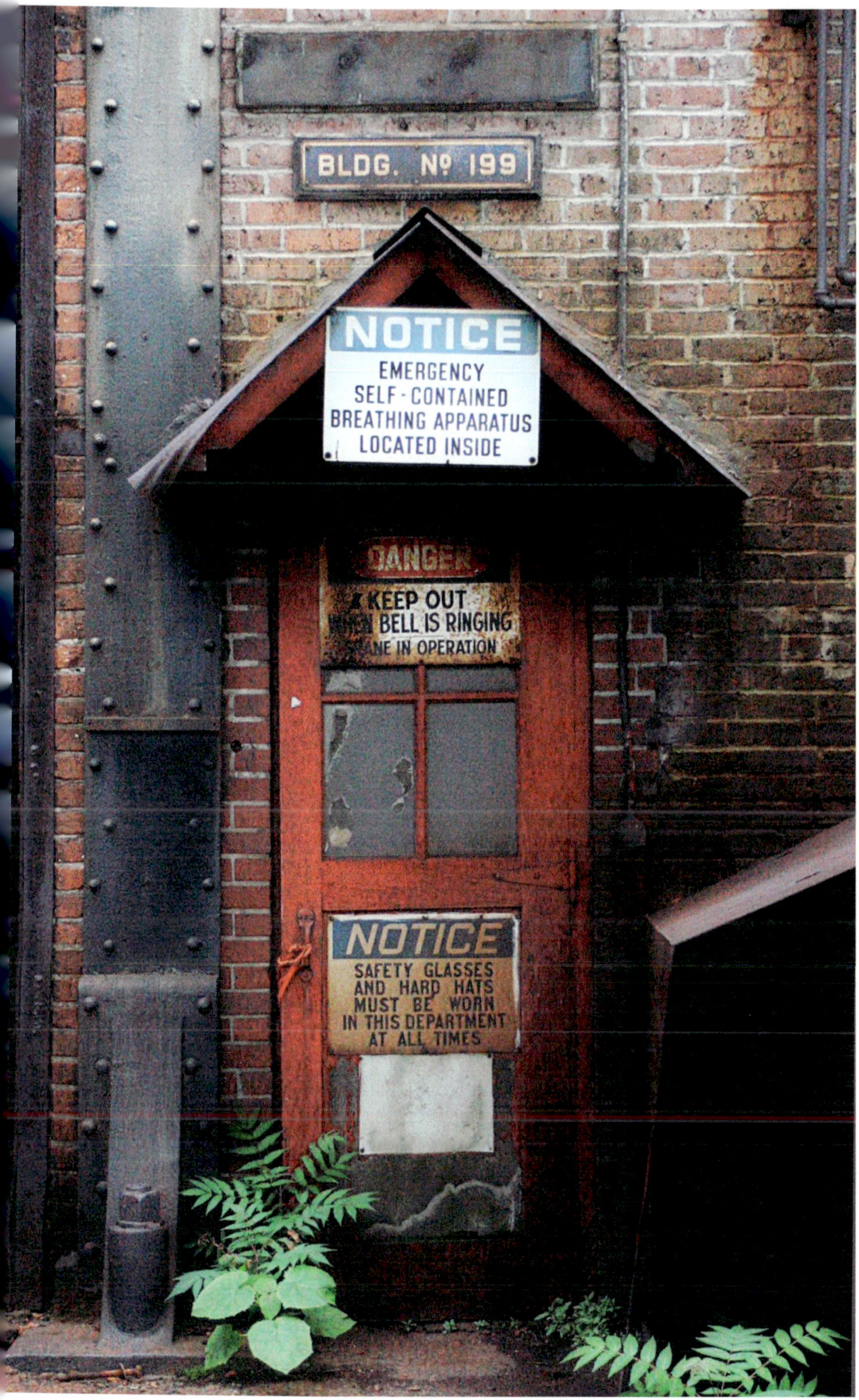
BLDG. Nº 199
NOTICE
EMERGENCY
SELF-CONTAINED
BREATHING APPARATUS
LOCATED INSIDE
KEEP OUT
BELL IS RINGING
IN OPERATION
NOTICE
SAFETY GLASSES
AND HARD HATS
MUST BE WORN
IN THIS DEPARTMENT
AT ALL TIMES

Left Picture:

Entrance to Oil Room

Right Picture:

Interior of Plant Office

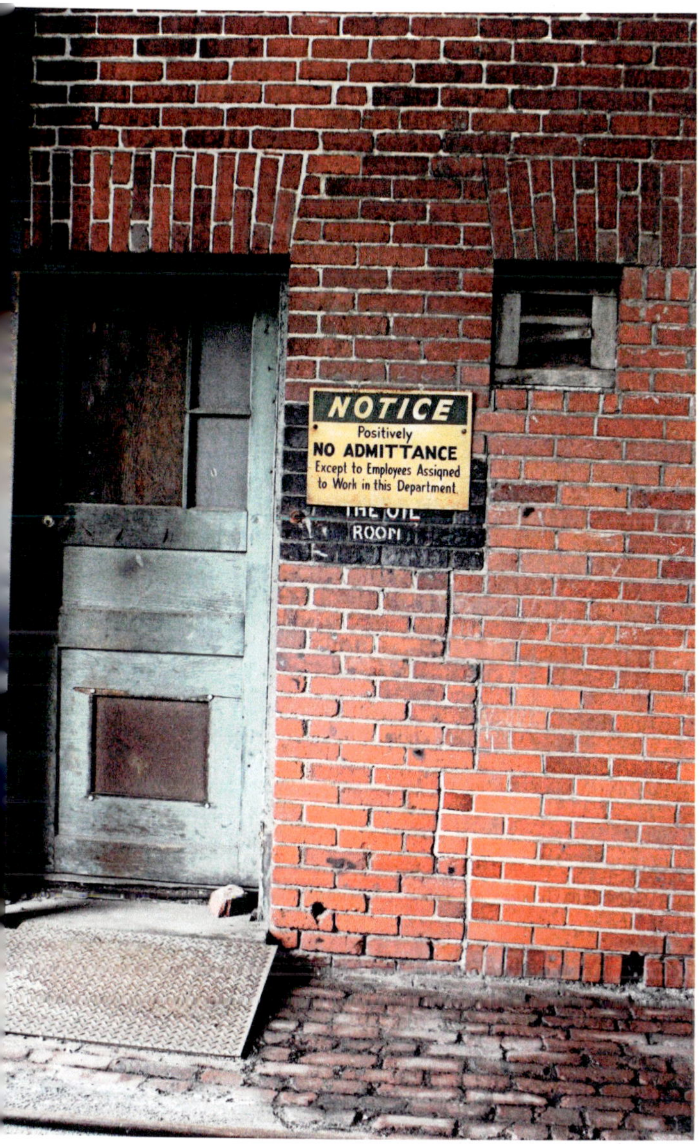
NOTICE
Positively
NO ADMITTANCE
Except to Employees Assigned
to Work in this Department
ROOM

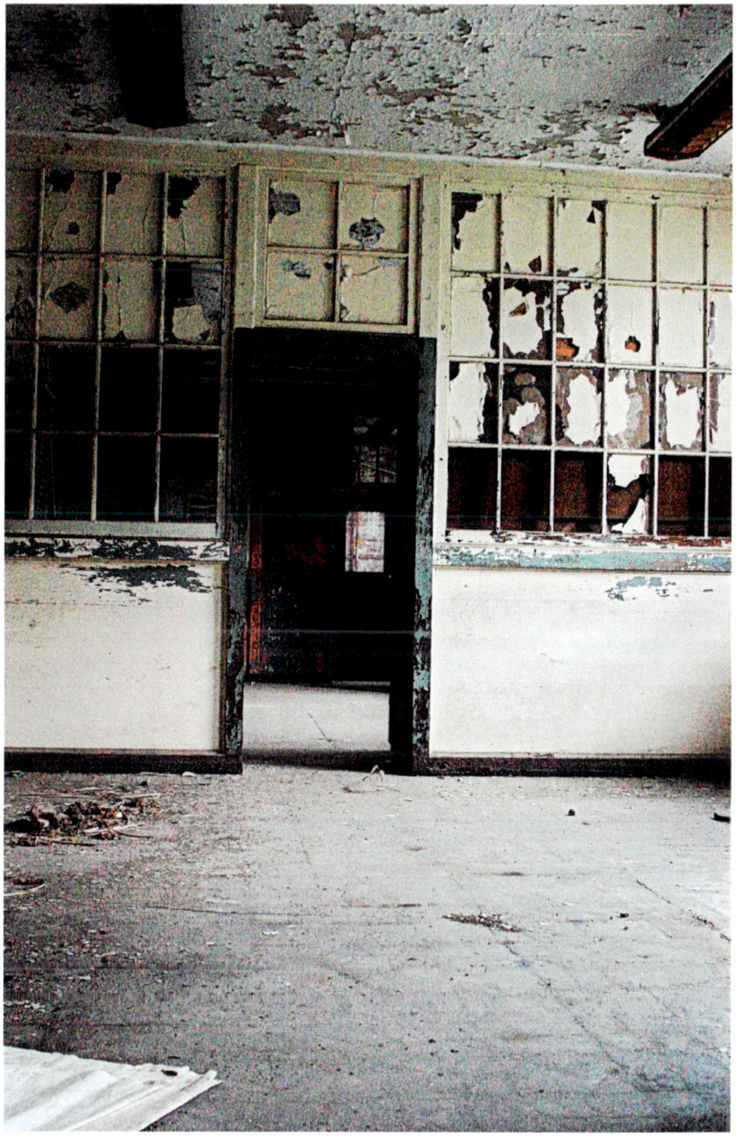

Bottom Picture:

Hoist mechanism associated with crane in Building 199 (No. 3 Vertical Treatment Building); 1920's.

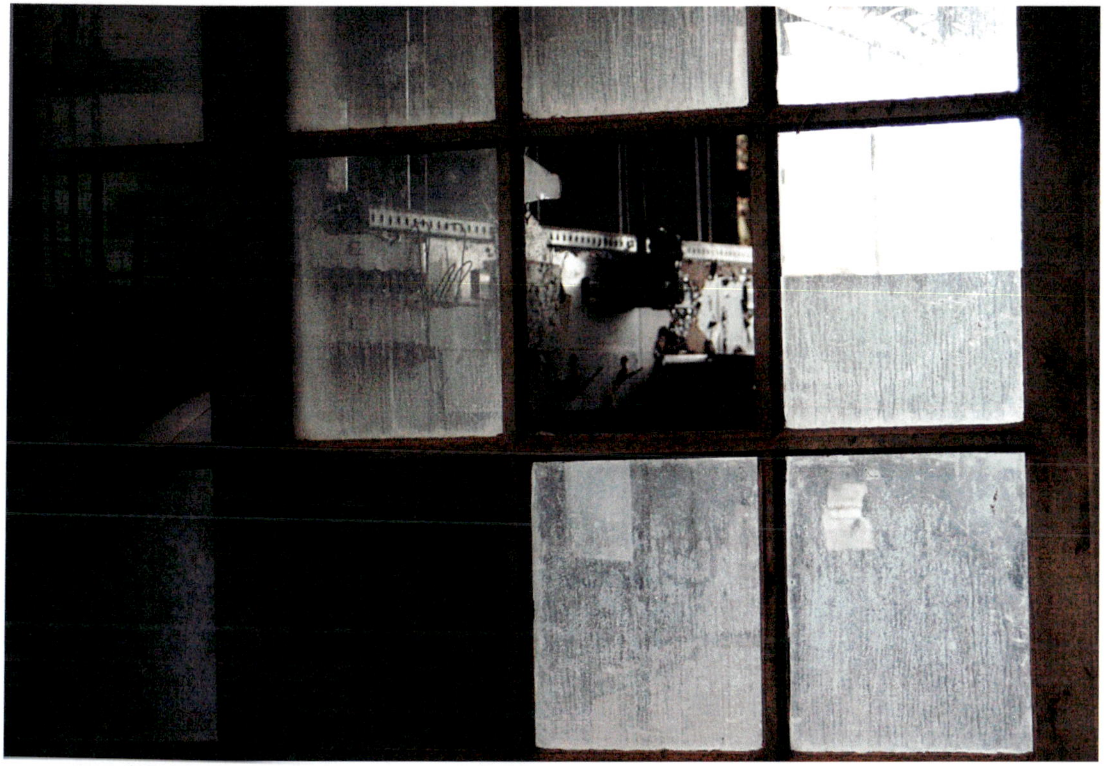

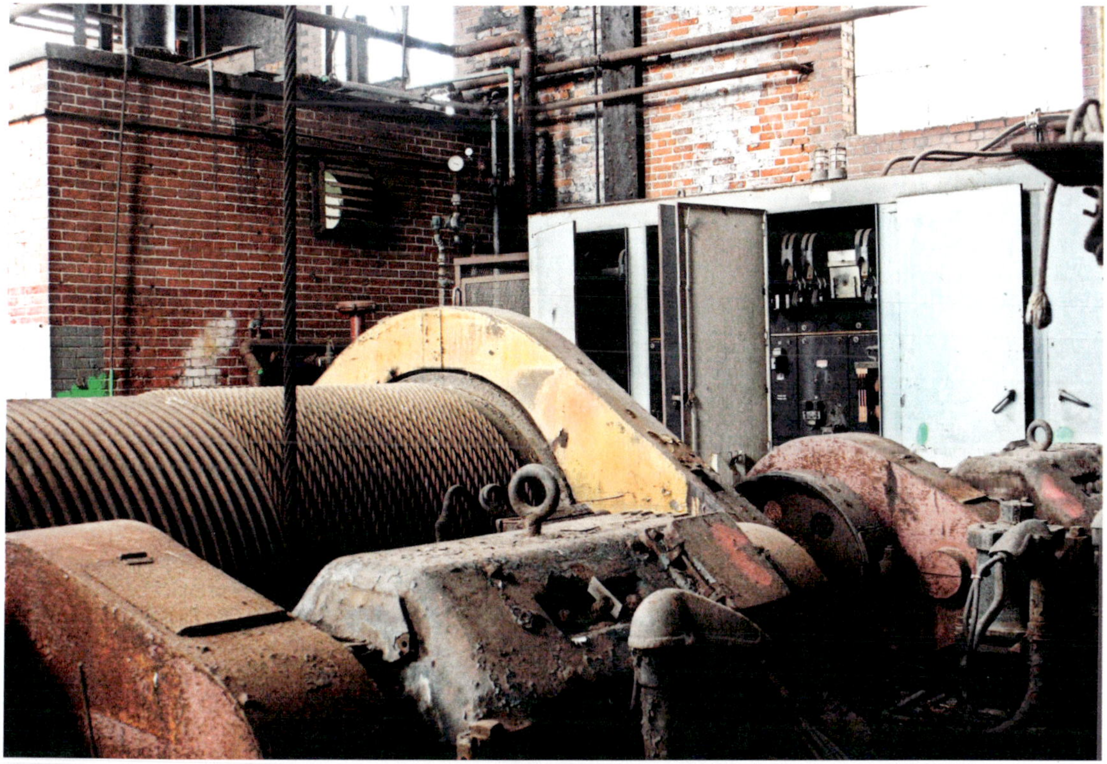

Various components of the crane in the
Number 3 Vertical Treatment Facility.

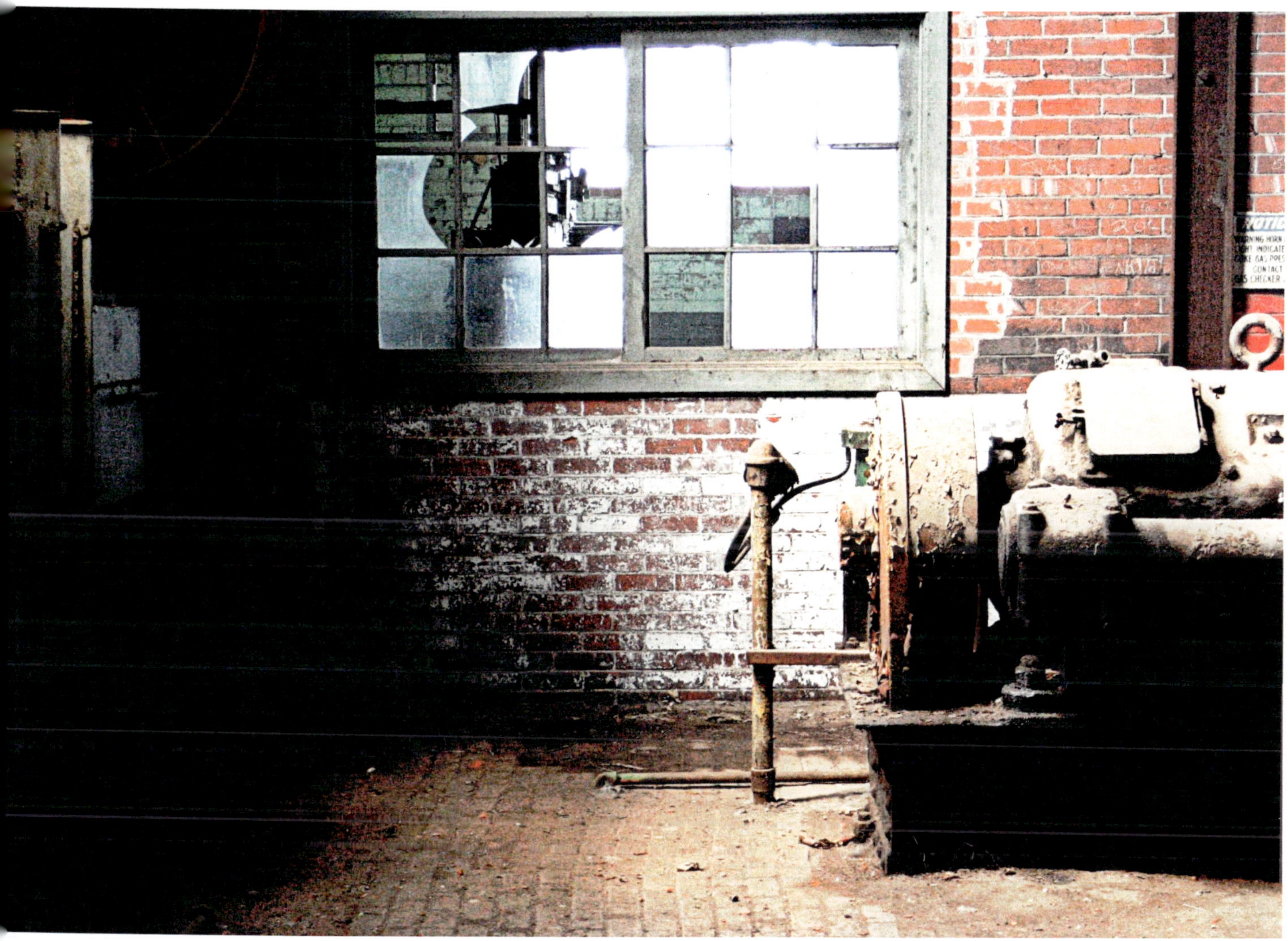

Original Bessemer Building circa 1872

Corner of the Original Bessemer Building; built 1872; Housed Bessemer converters.

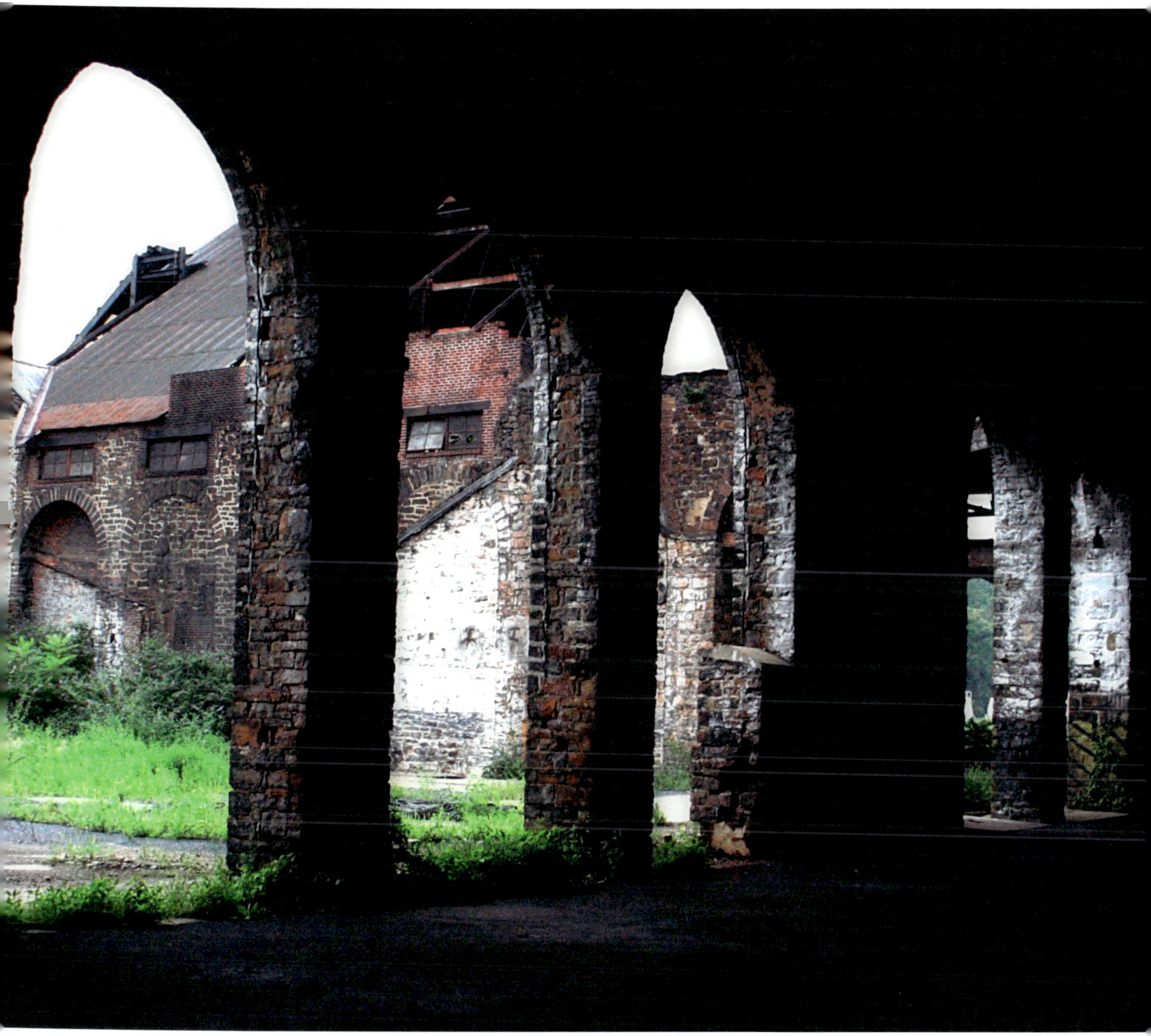

Original Bessemer Building circa 1872

Bessemer Building; steel rolling facility and then was the iron foundry; Western end of the Iron Foundry also known as the Bessemer Building.

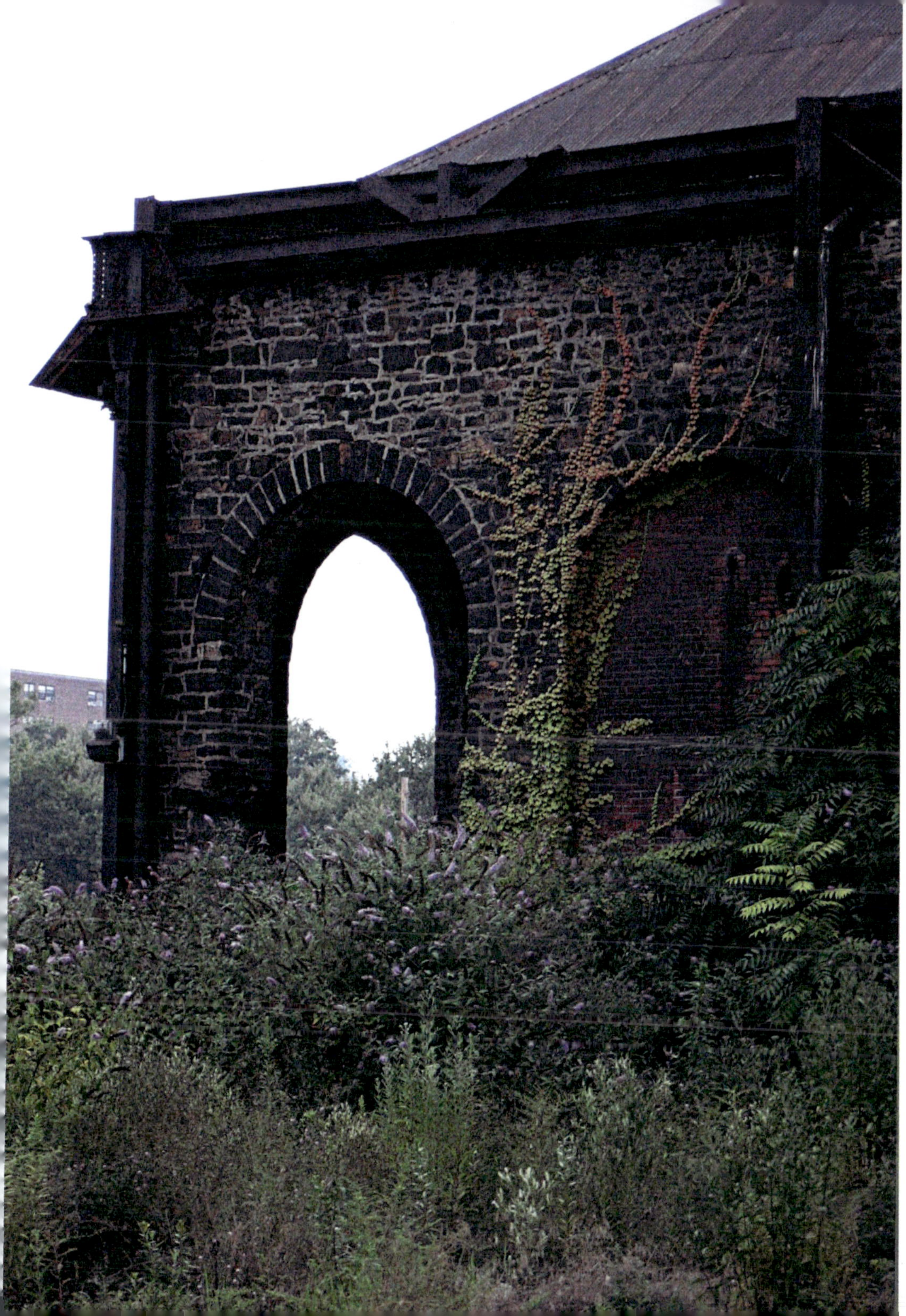

Top Picture:

Truss work in Original Bessemer Building.

Bottom Picture:

Electrical safety switches mounted on a wall inside the Bethlehem Plant.

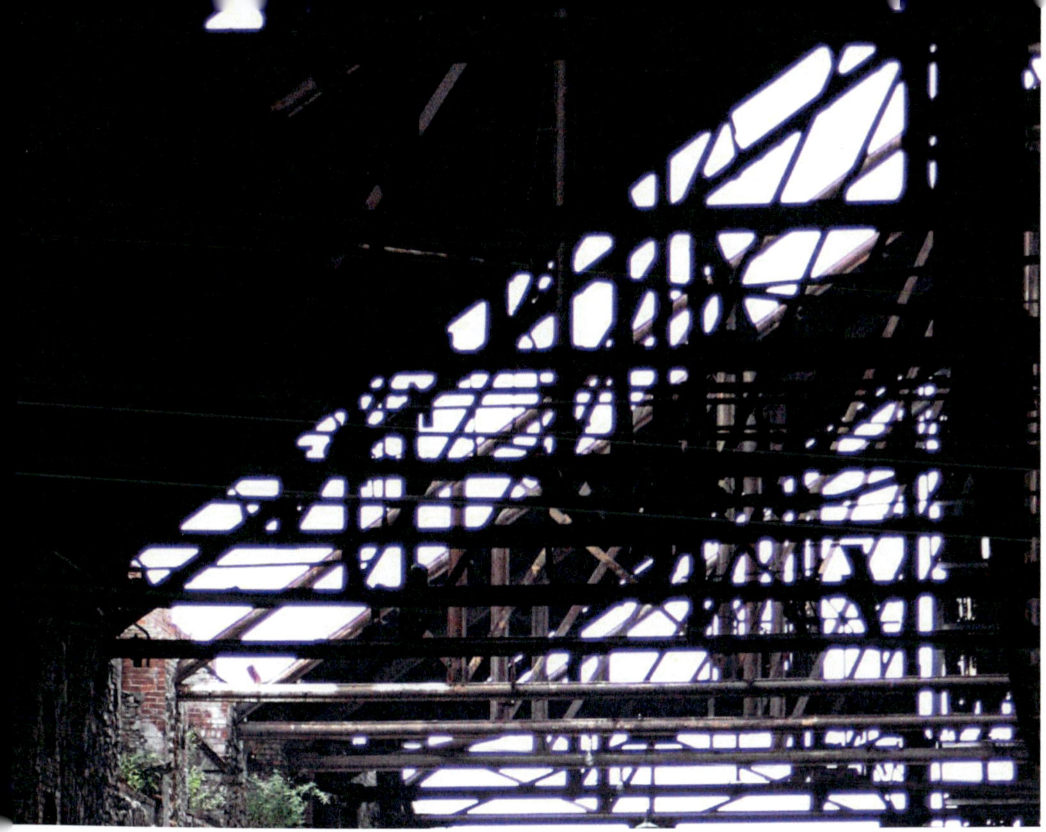

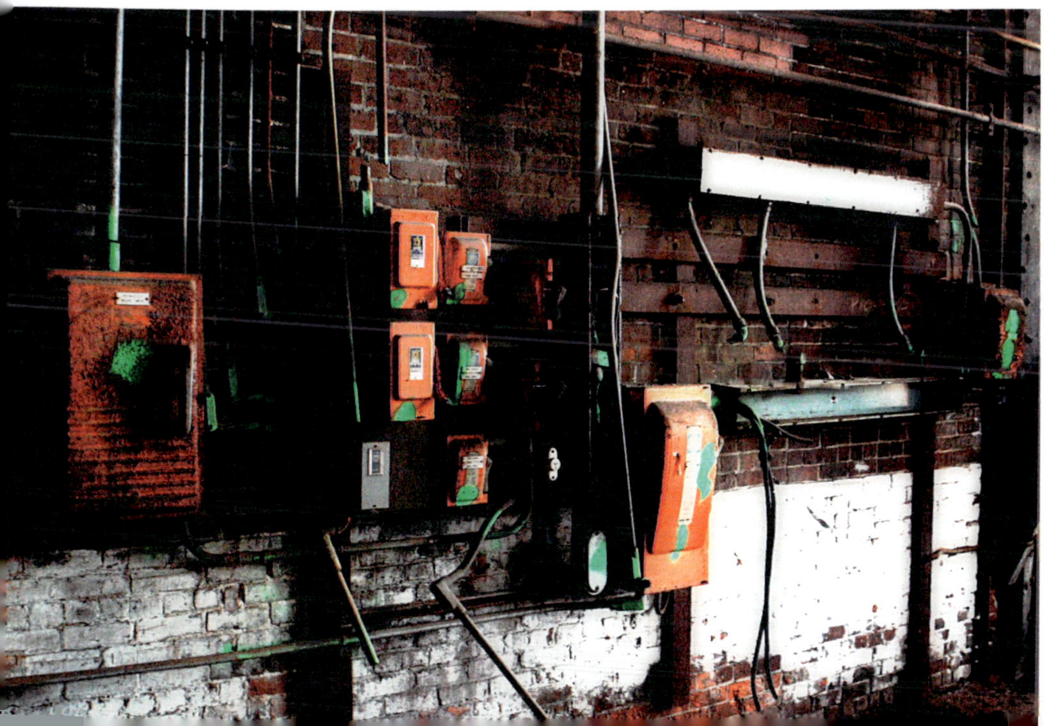

Floor of High House Interior

Vertical Furnace Inside No. 3 Vertical treatment Building; look past furnace into Welfare Room.

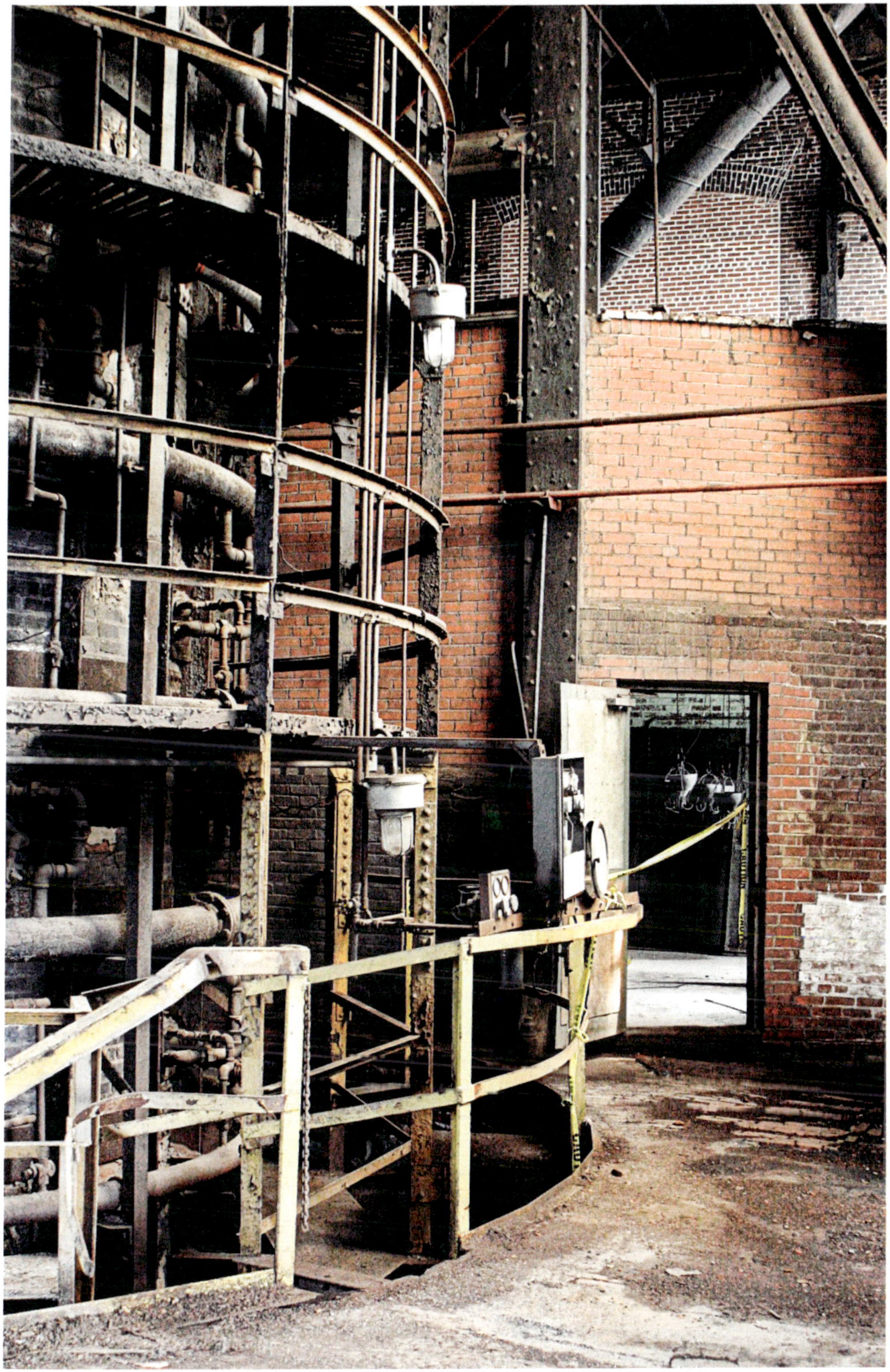

Meter

Early nomadic instrumentation for controlling a production process inside the Bethlehem Plant.

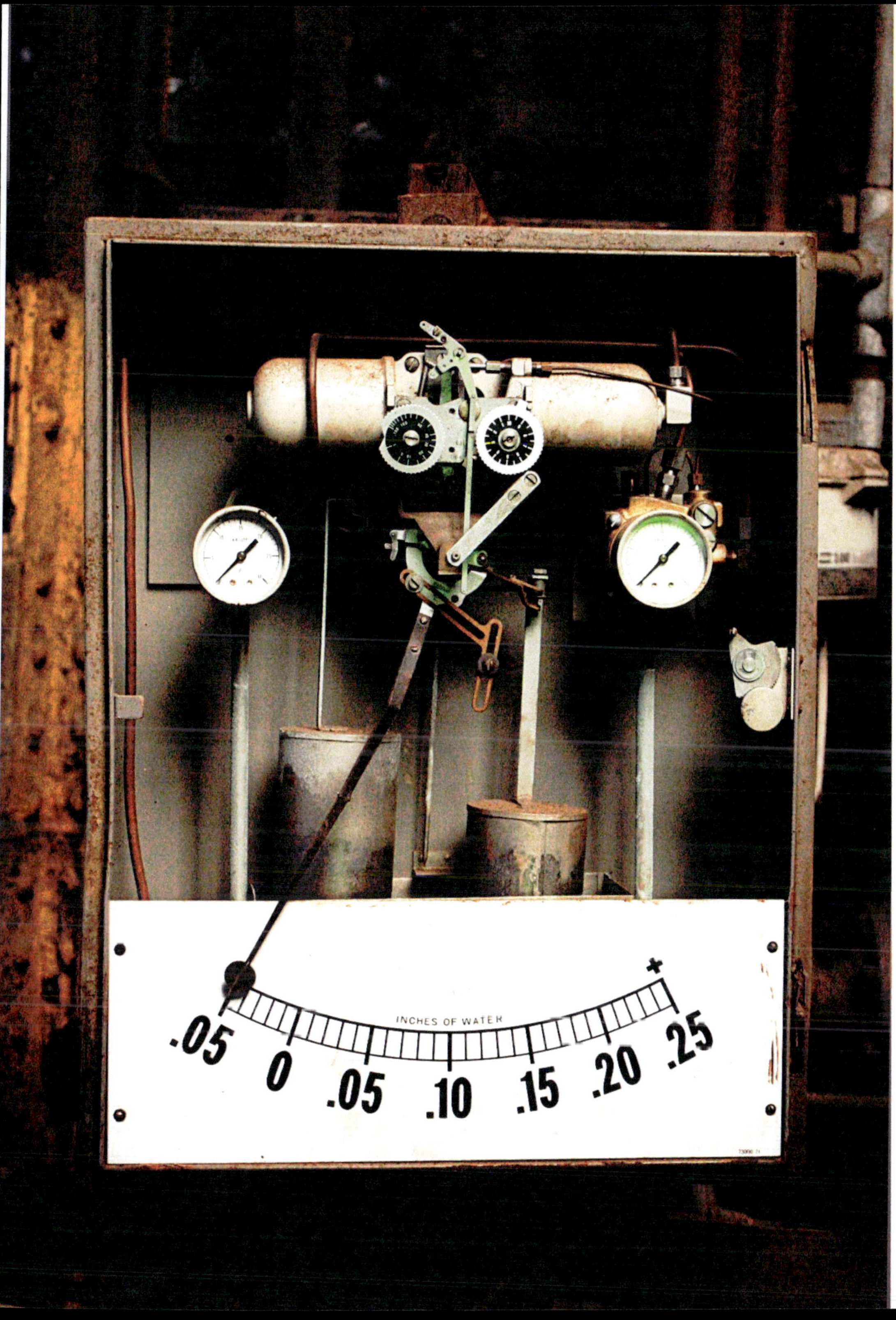
INCHES OF WATER
.05 0 .05 .10 .15 .20 .25
+

Left Picture:

Truss work & Blast Furnace.

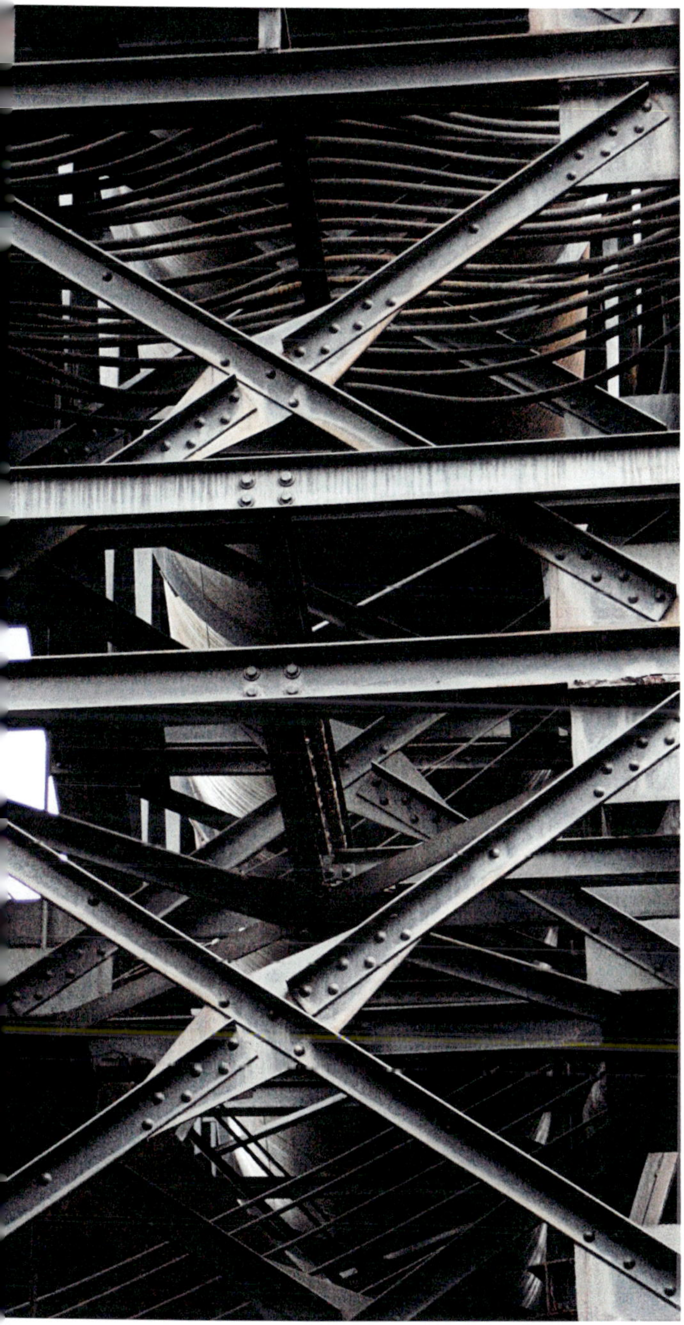

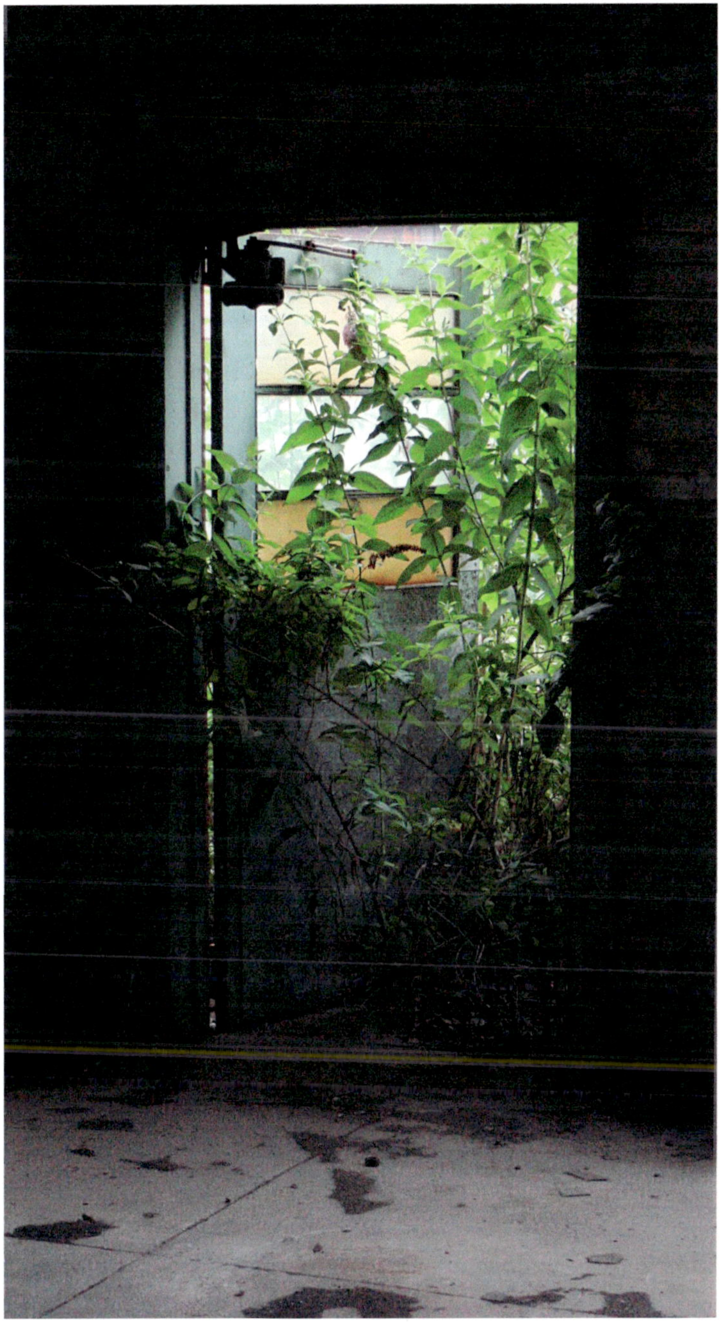

Blast Furnace

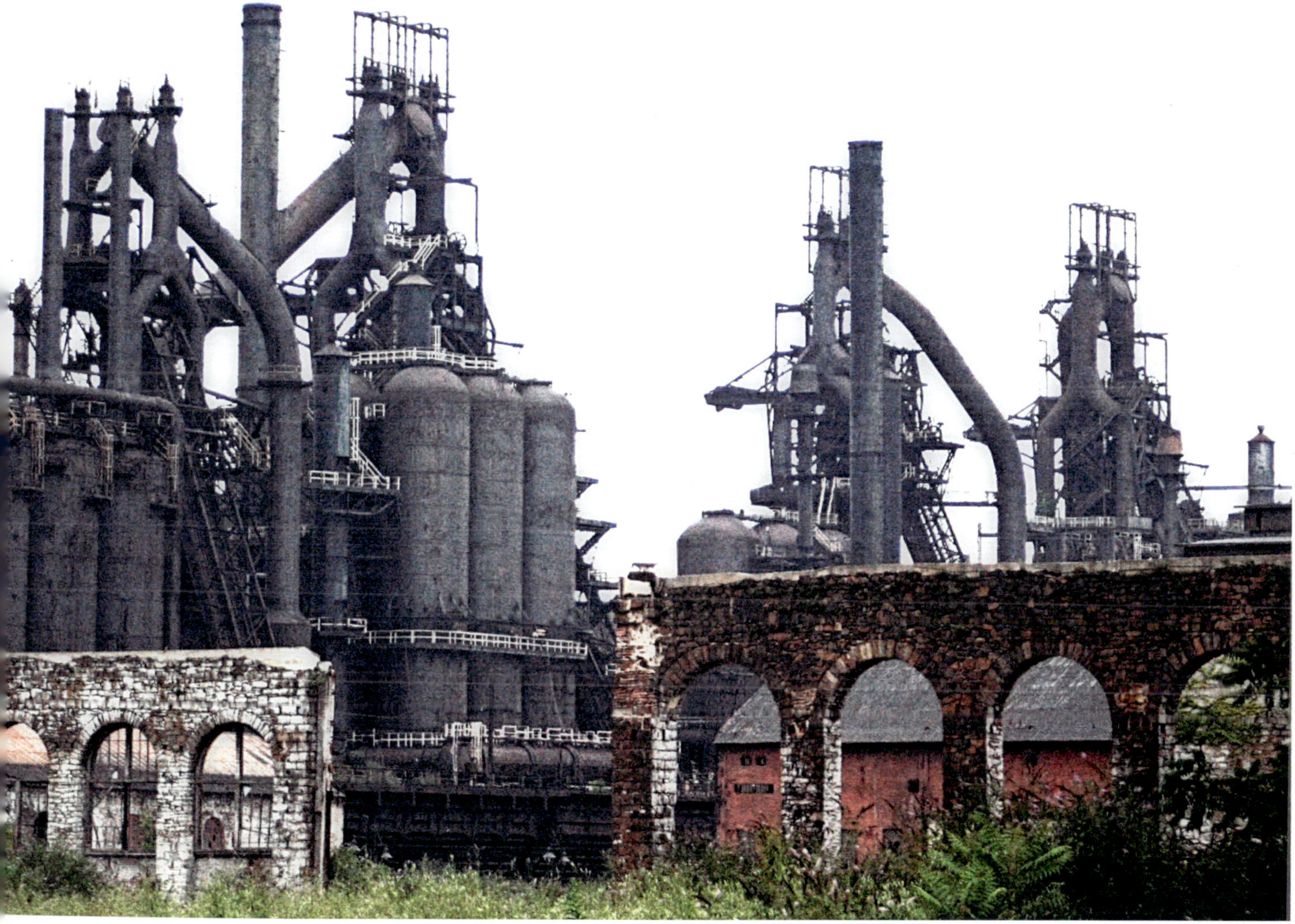

Carpenter Shop

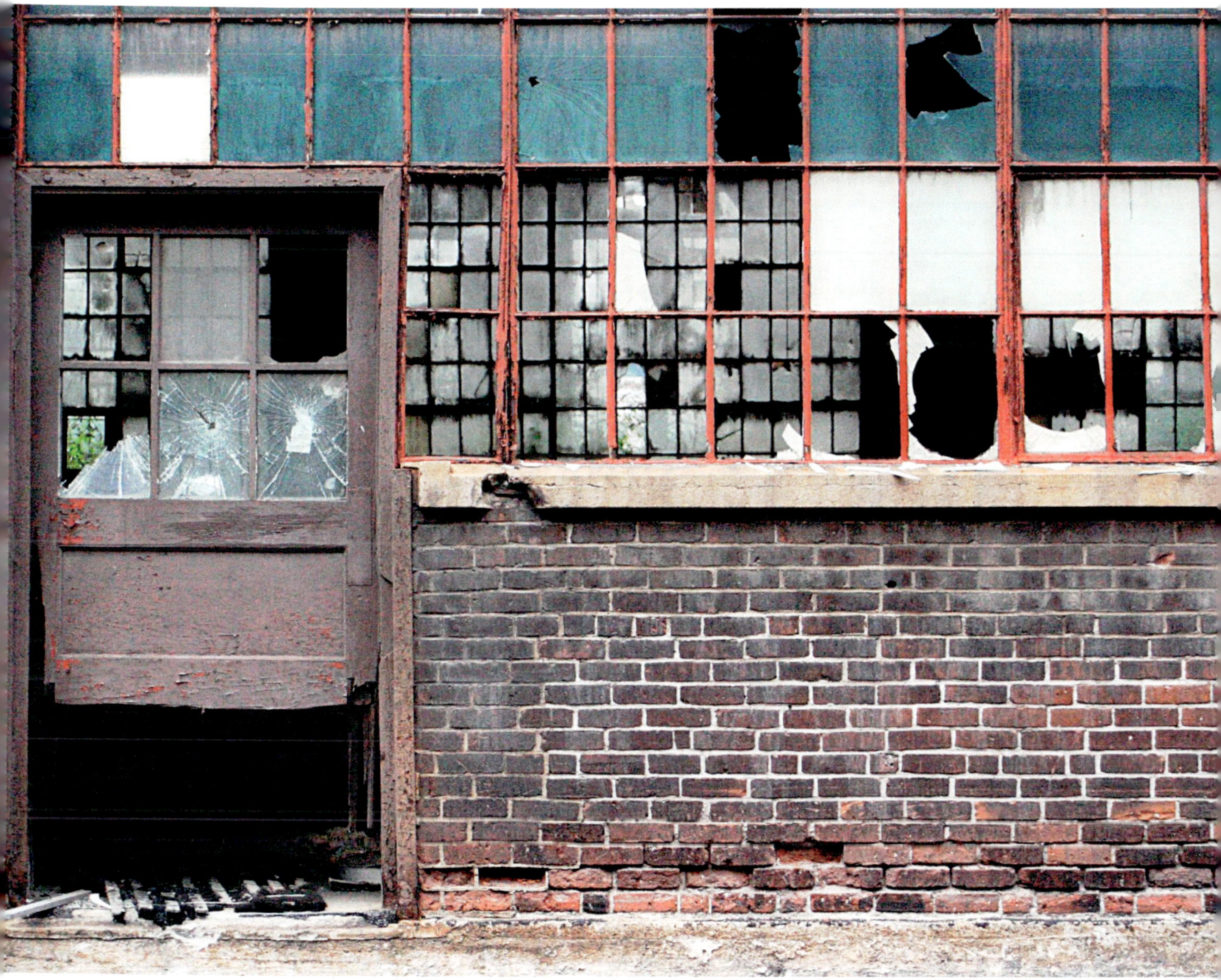

Roof cast

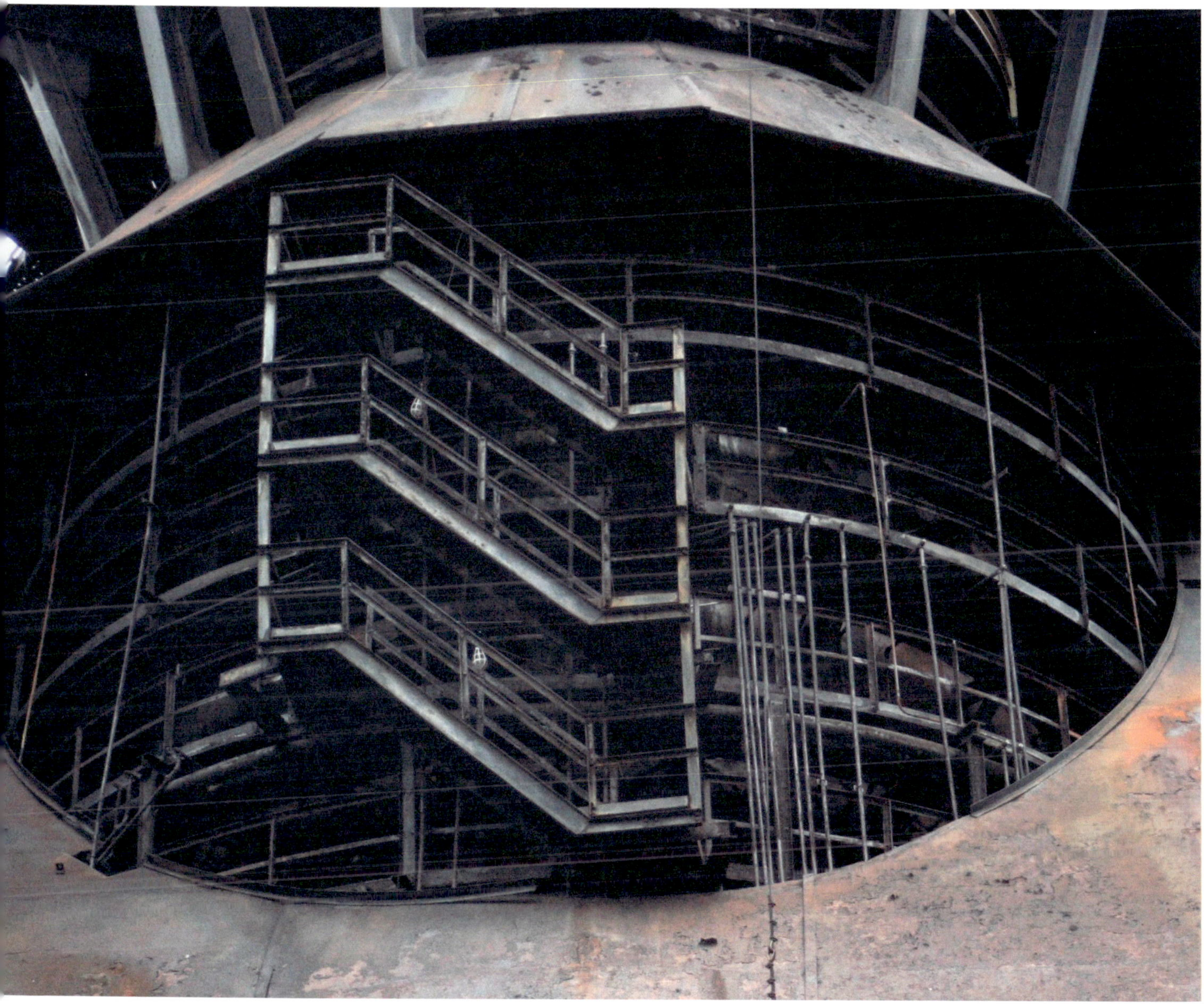

Interior of Carpenter Shop

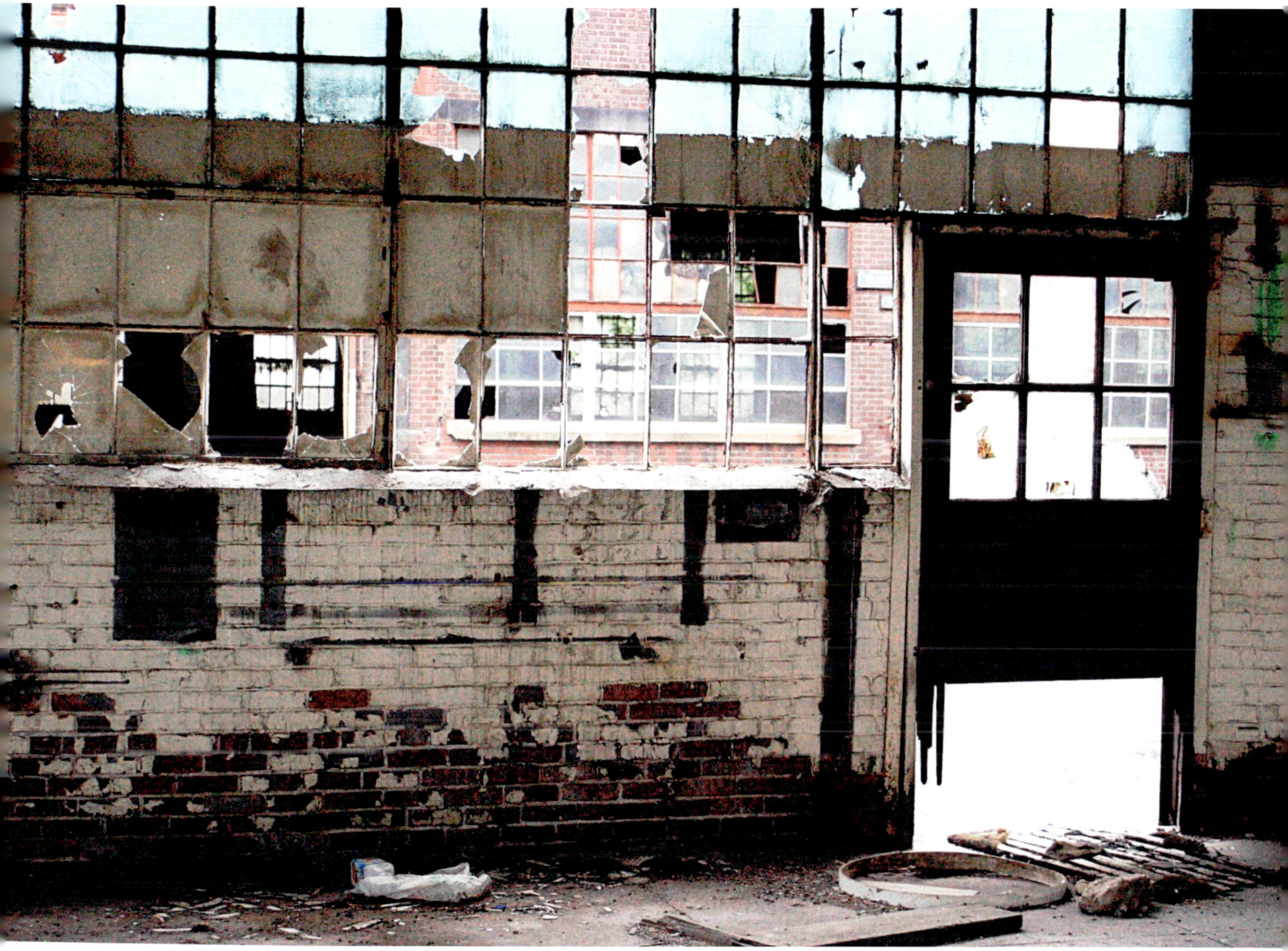

Interior of Carpenter Shop

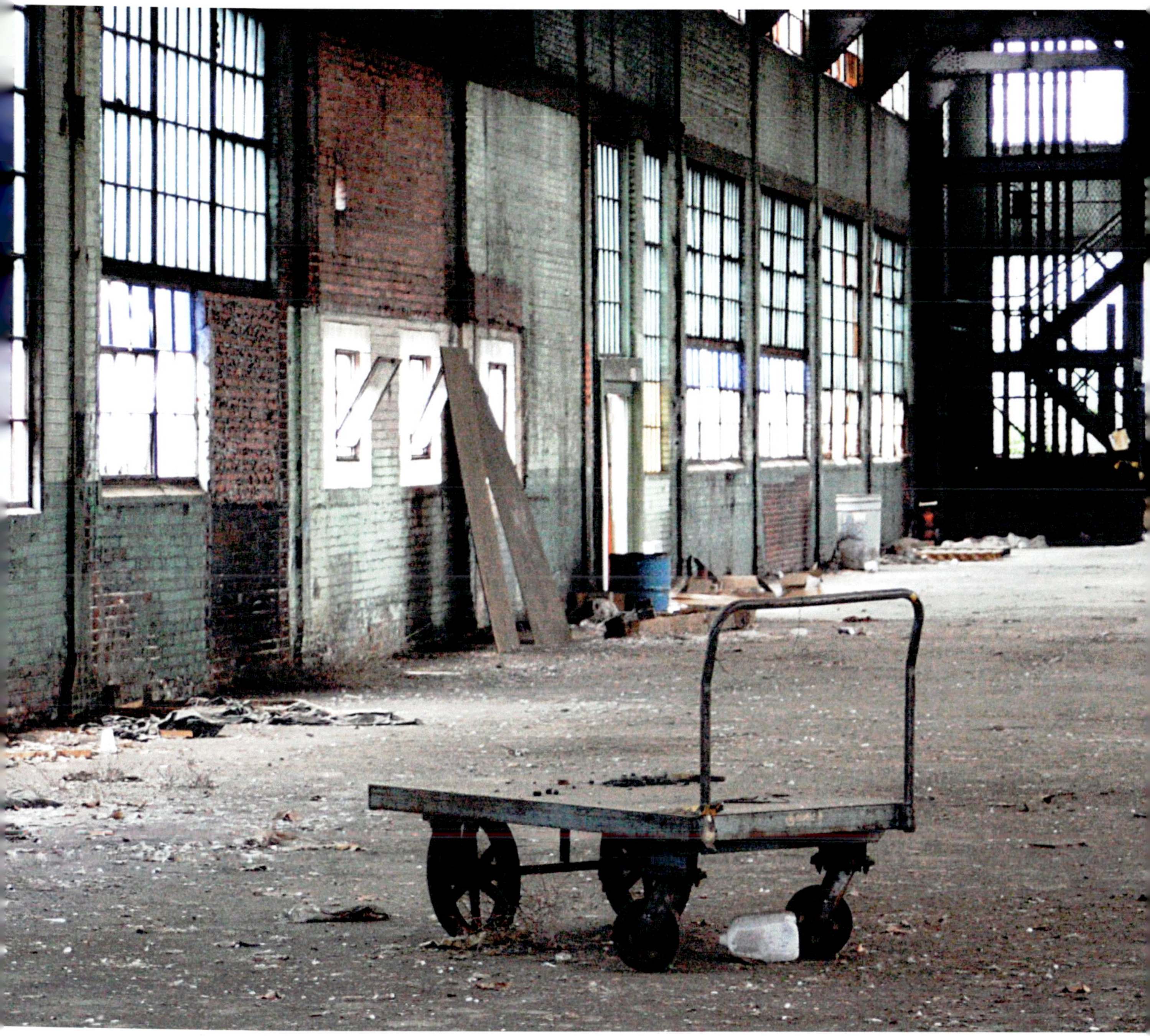

Bottom Picture:

Remains of Original Bessemer Steel Rail Facility.

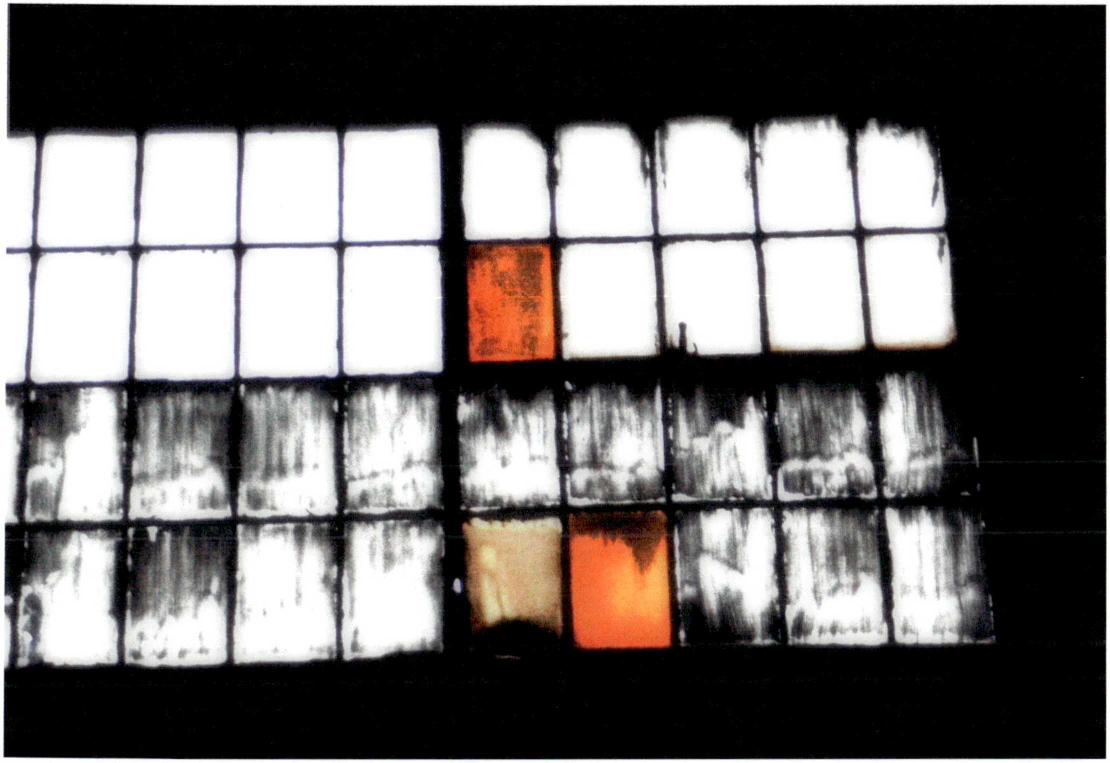

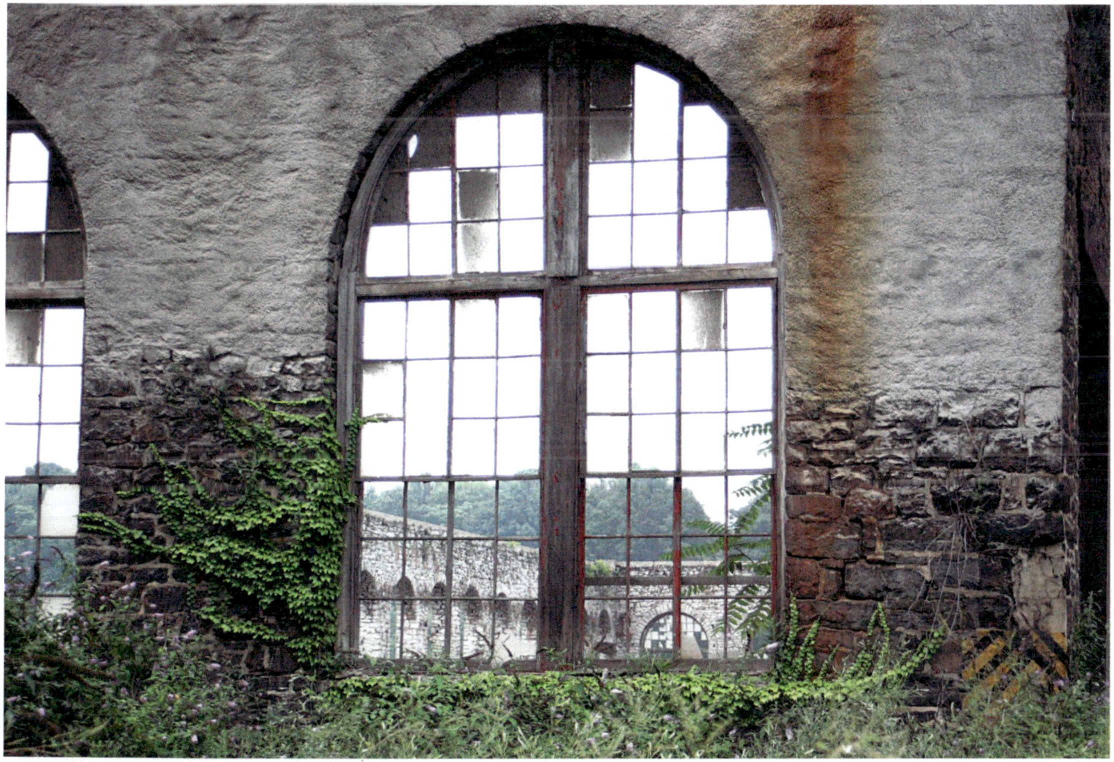

Controls used to keep pipes from freezing.

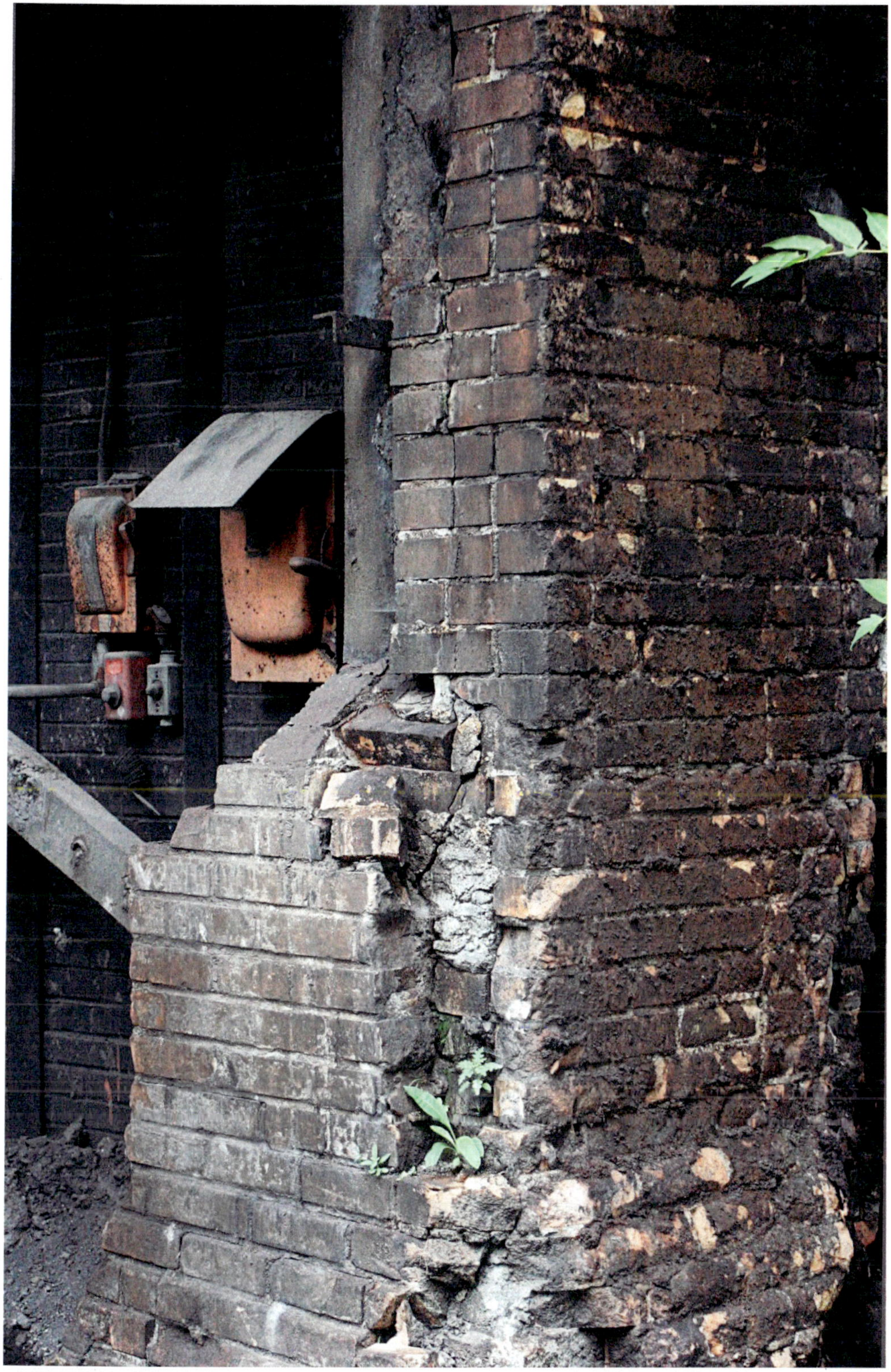

Interior of Electric Repair Shop

Southeast corner of electrical repair shop; ex-elevator shaft; staircase going to the second floor of electrical repair shop; staircase in southeast corner of electrical repair shop near blast furnace.

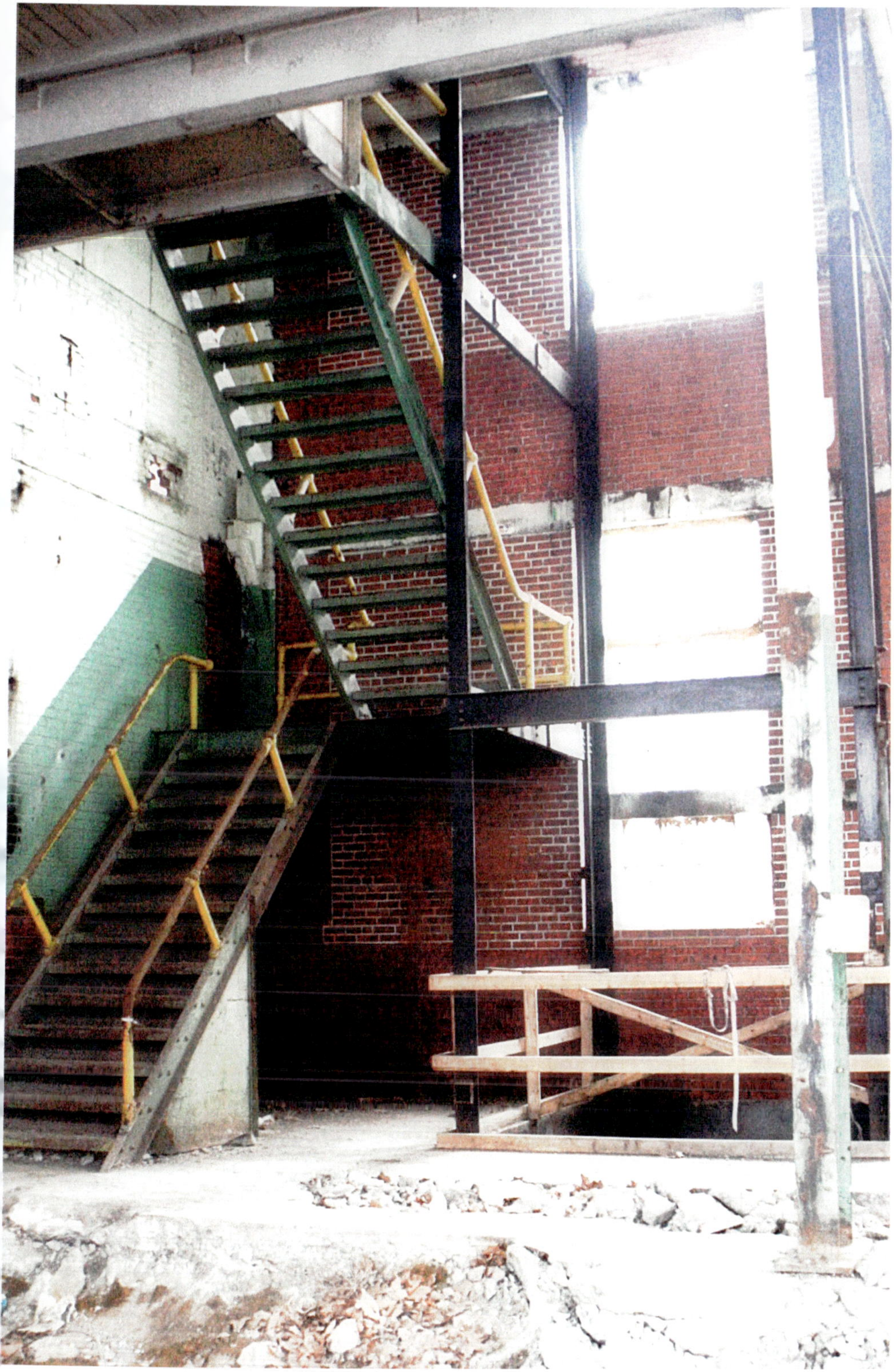

Top Picture:

Blast Furnace E

Bottom Picture:

Blast furnace area; near blast furnace E.

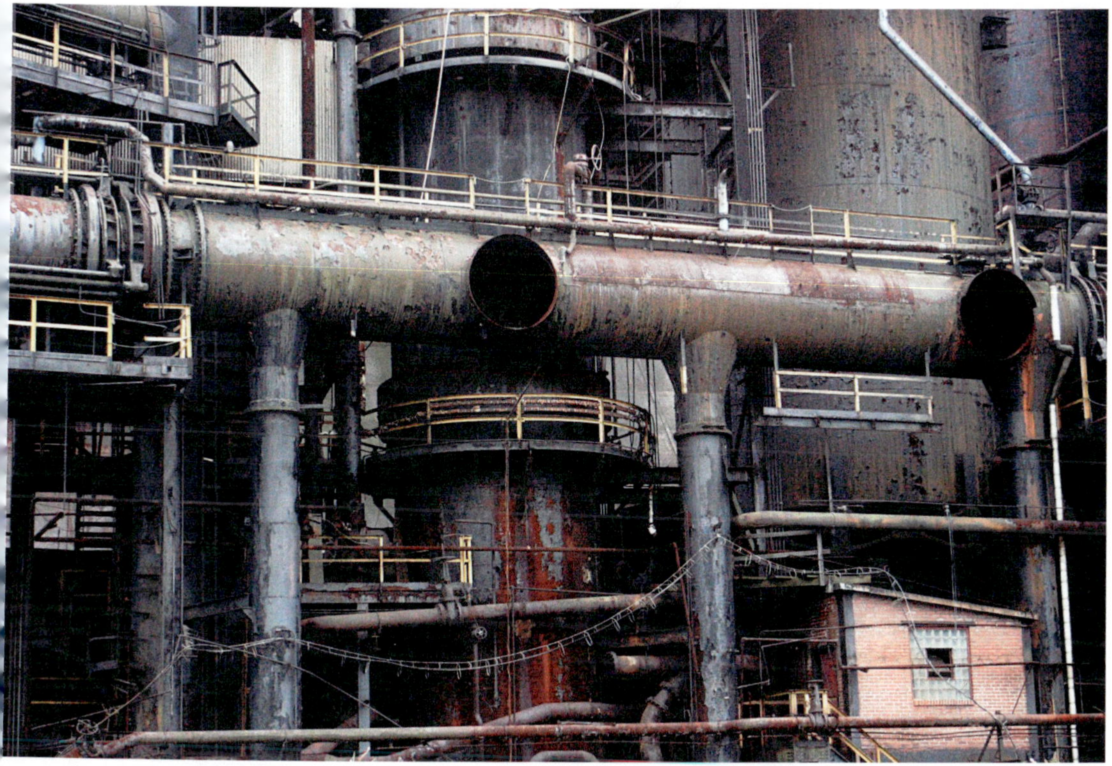

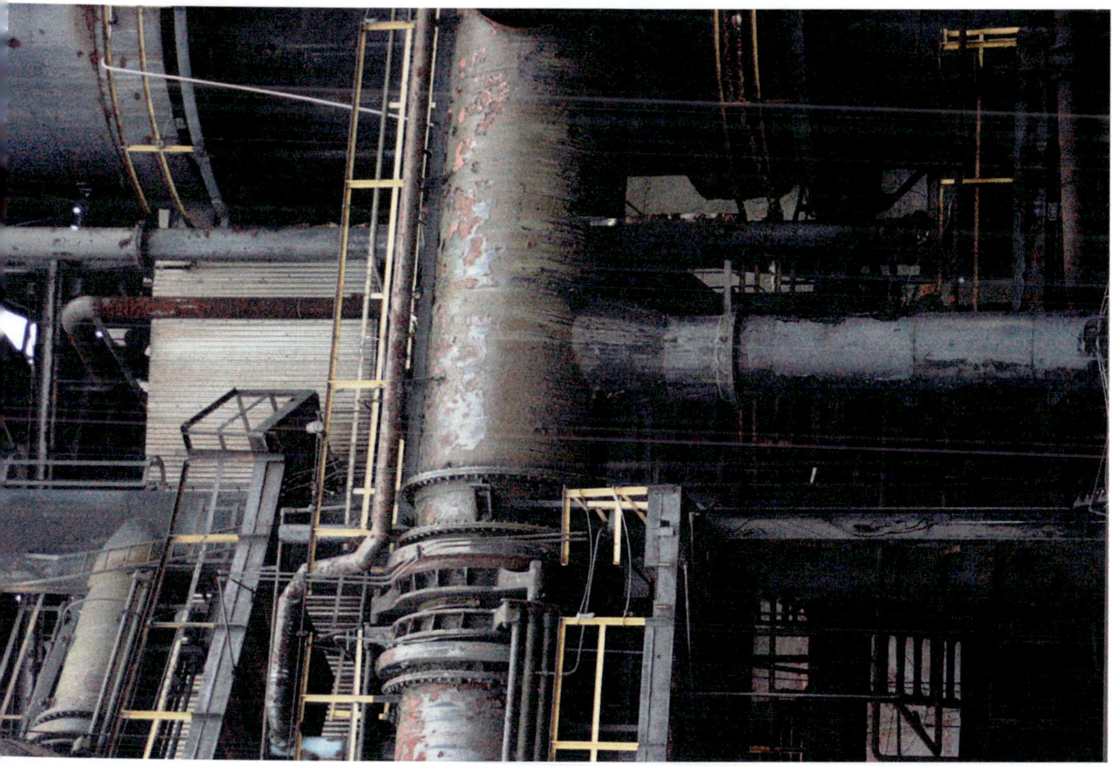

Blast Furnace

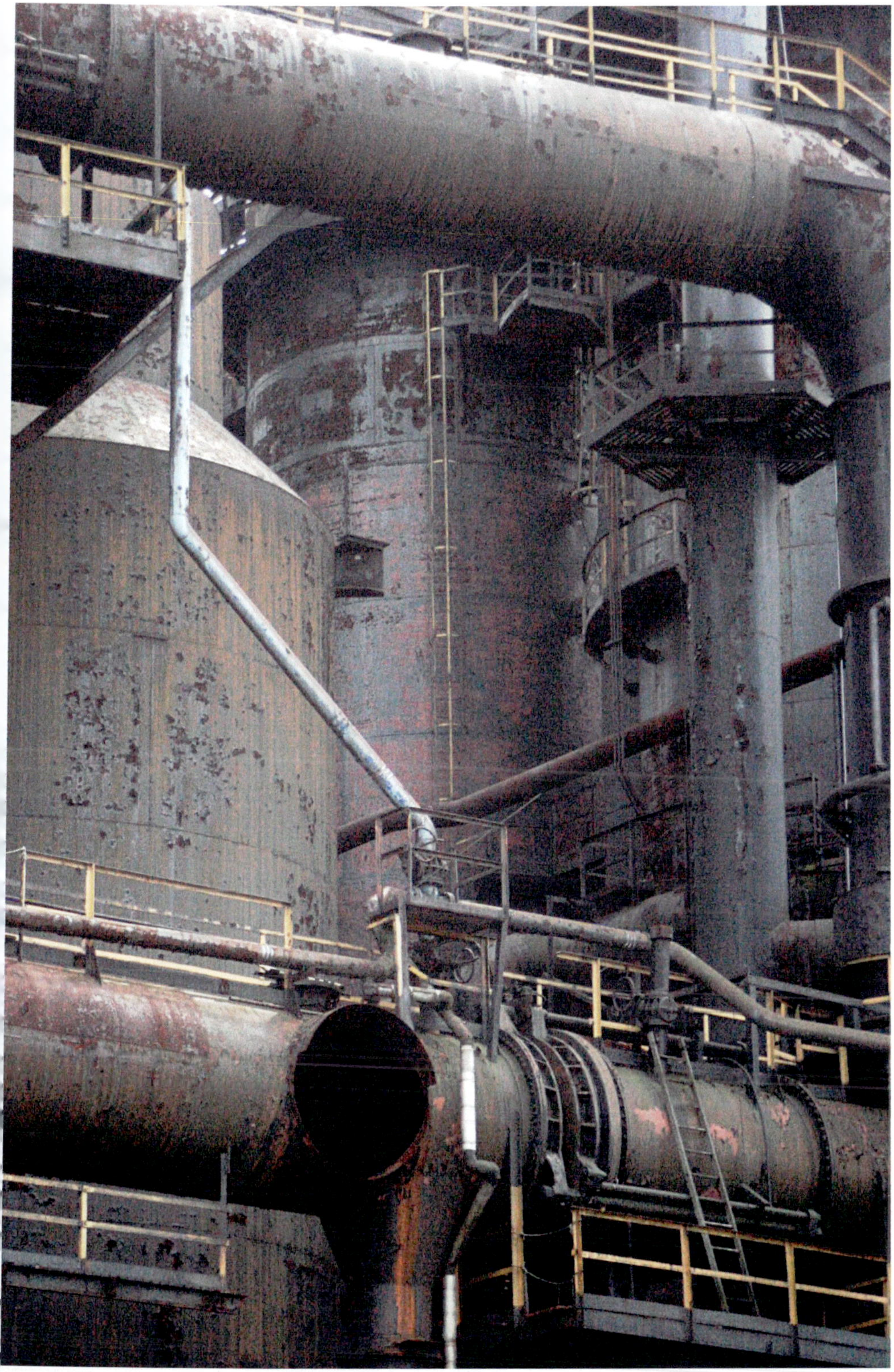

Top Picture:

Blast Furnace

Bottom Picture:

Each blast furnace had 4 stoves; only 3 for furnace C.

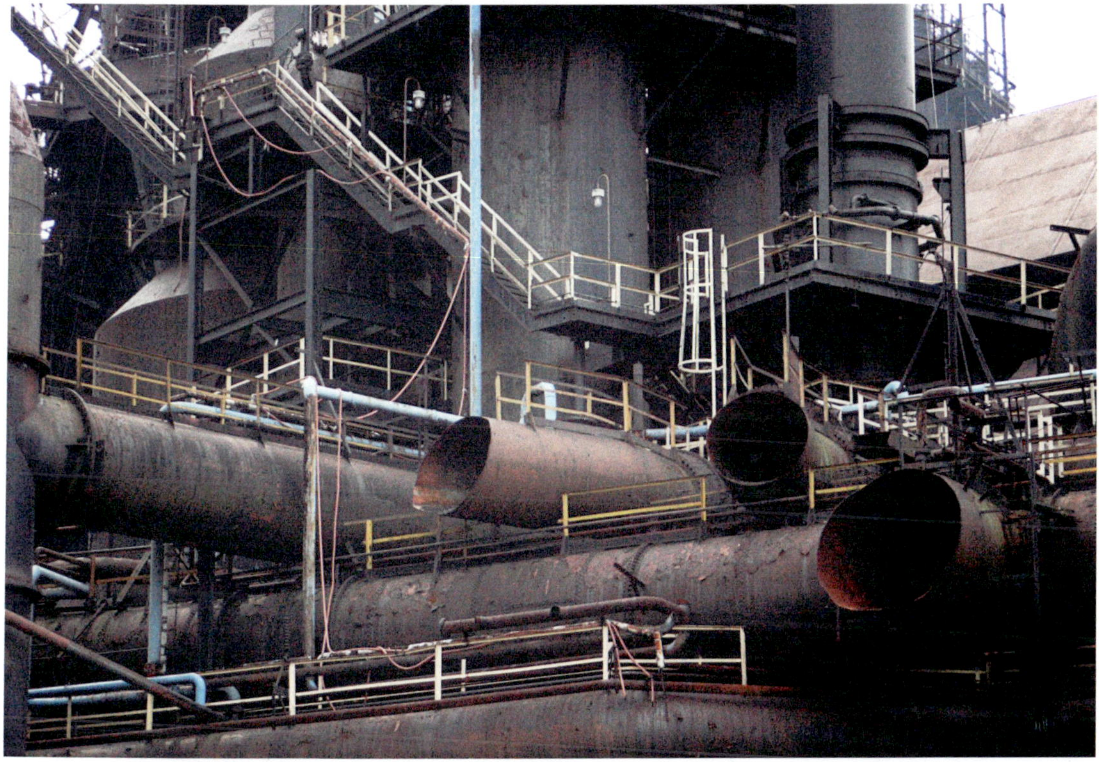

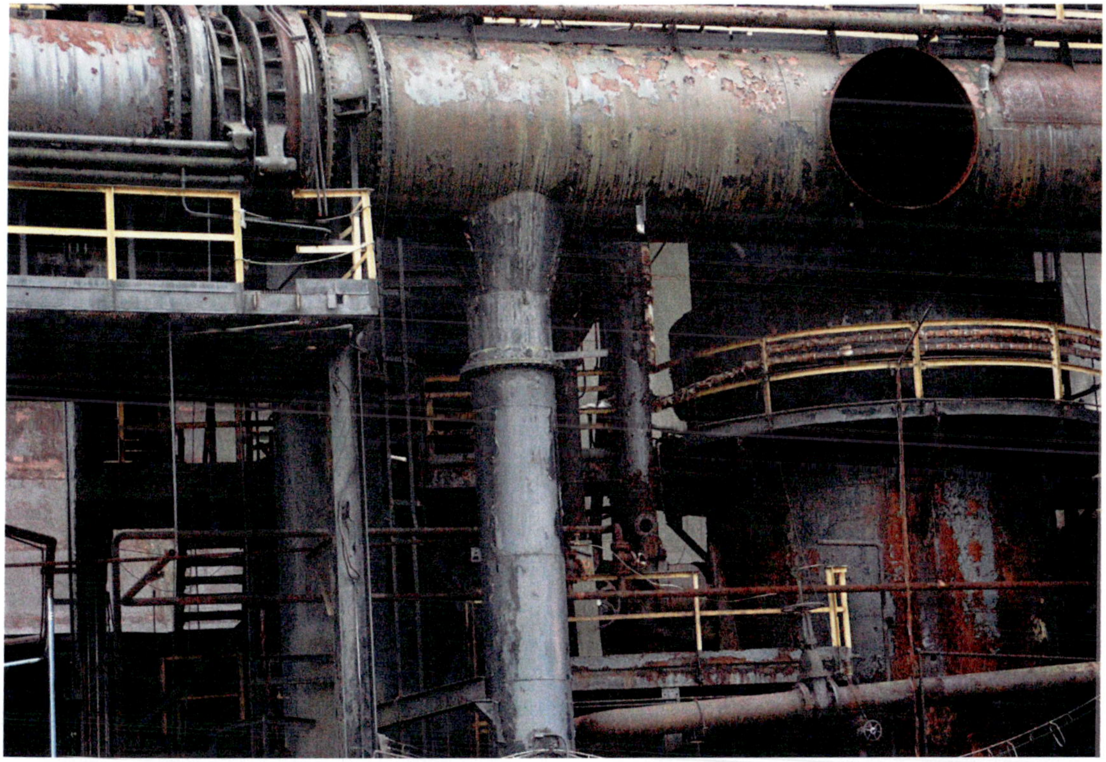

Control room for equipment in area.

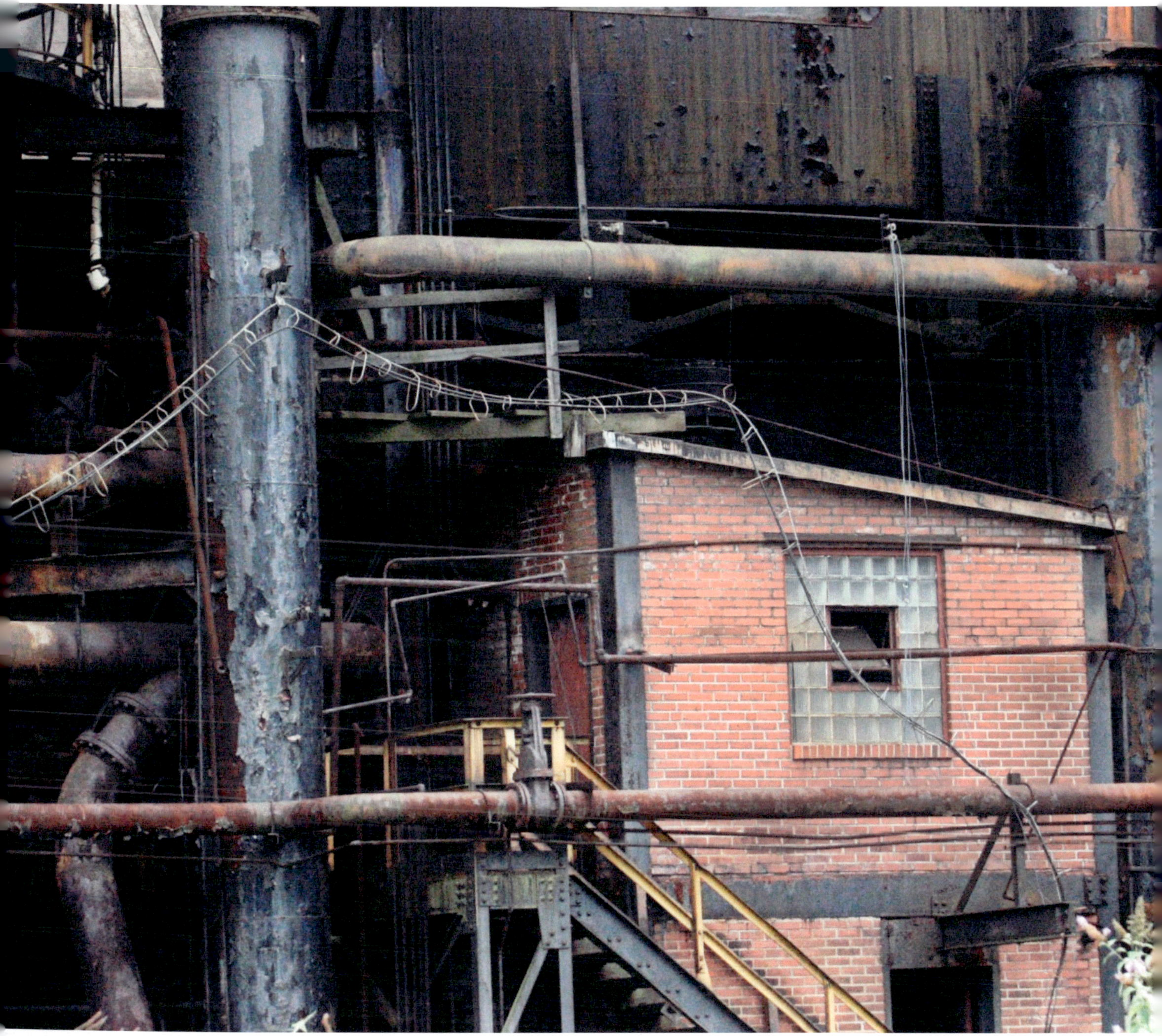

Gas Distribution System

Blast Furnace Detail of Gas Line

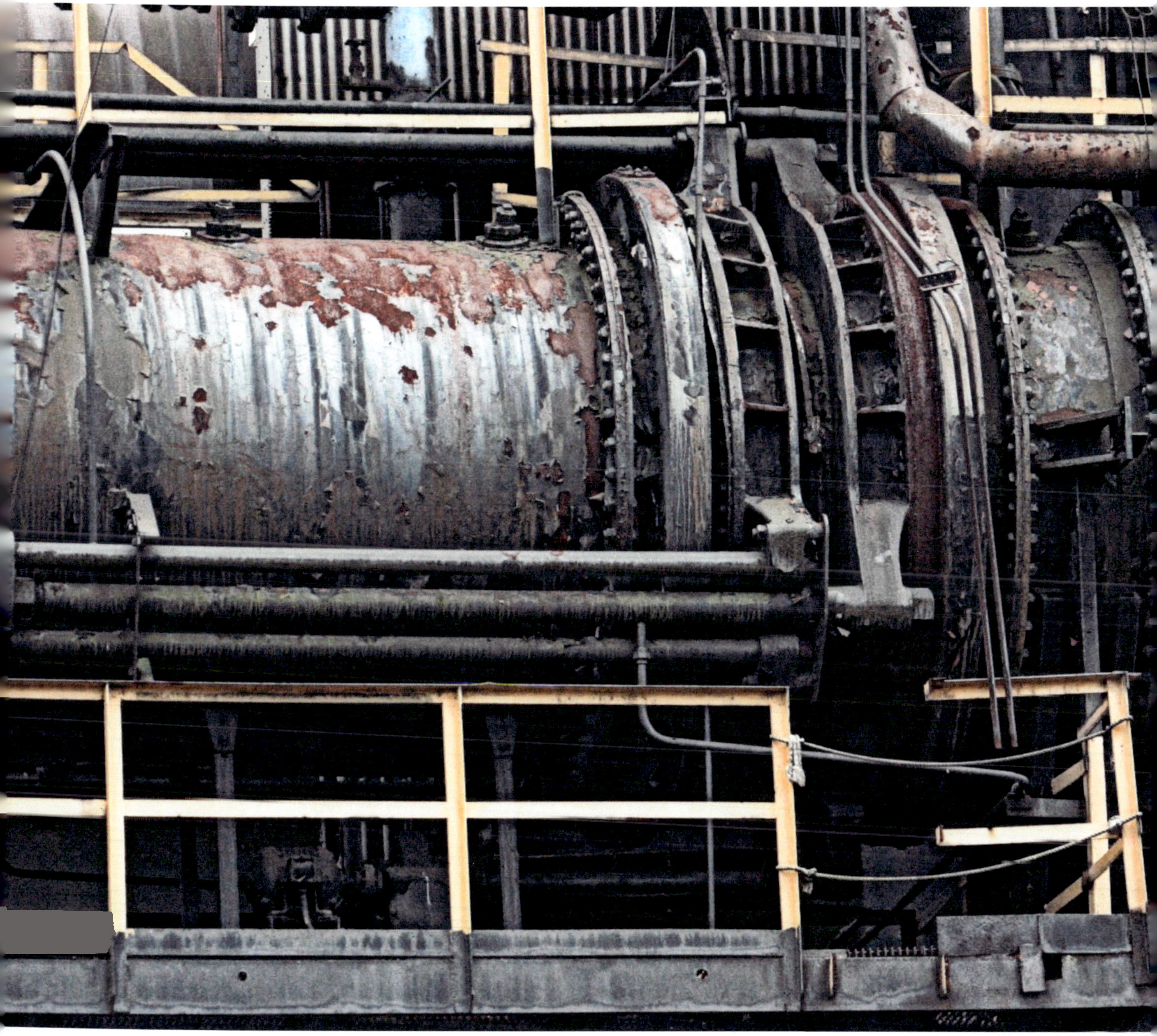

Blast Furnace

Blast furnace area; stairway accessing various components of gas cleaning system.

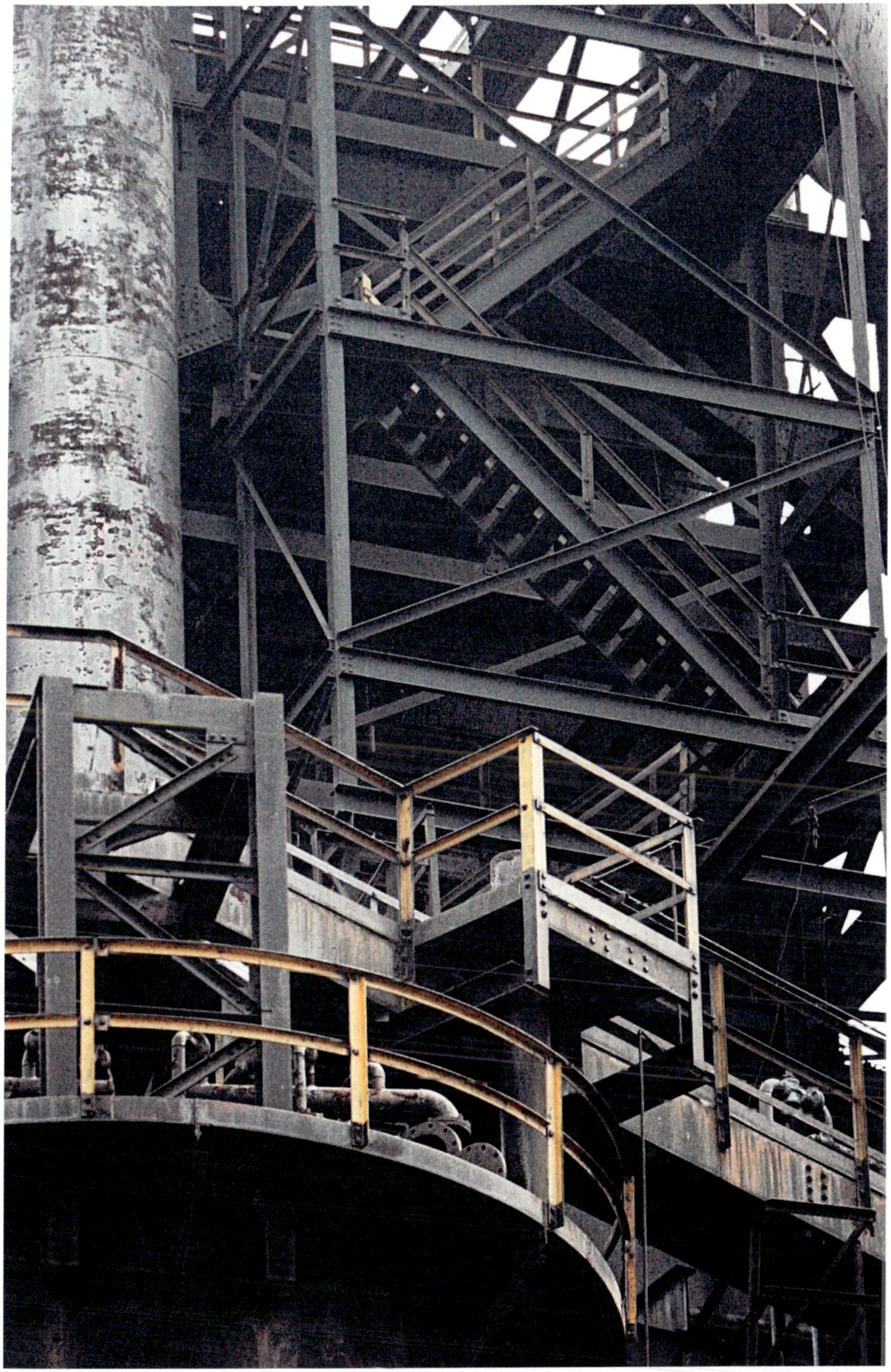

Left Picture:

Upper part of furnace (tree); charging floor; roof cast.

Right Picture:

Blast Furnace Office

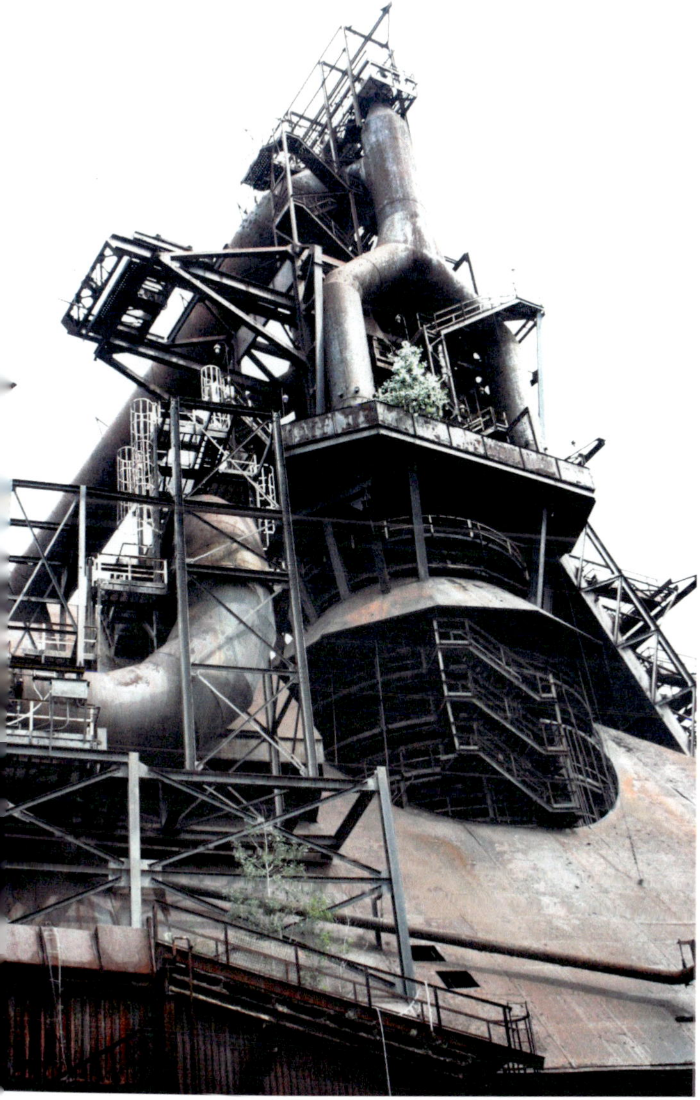

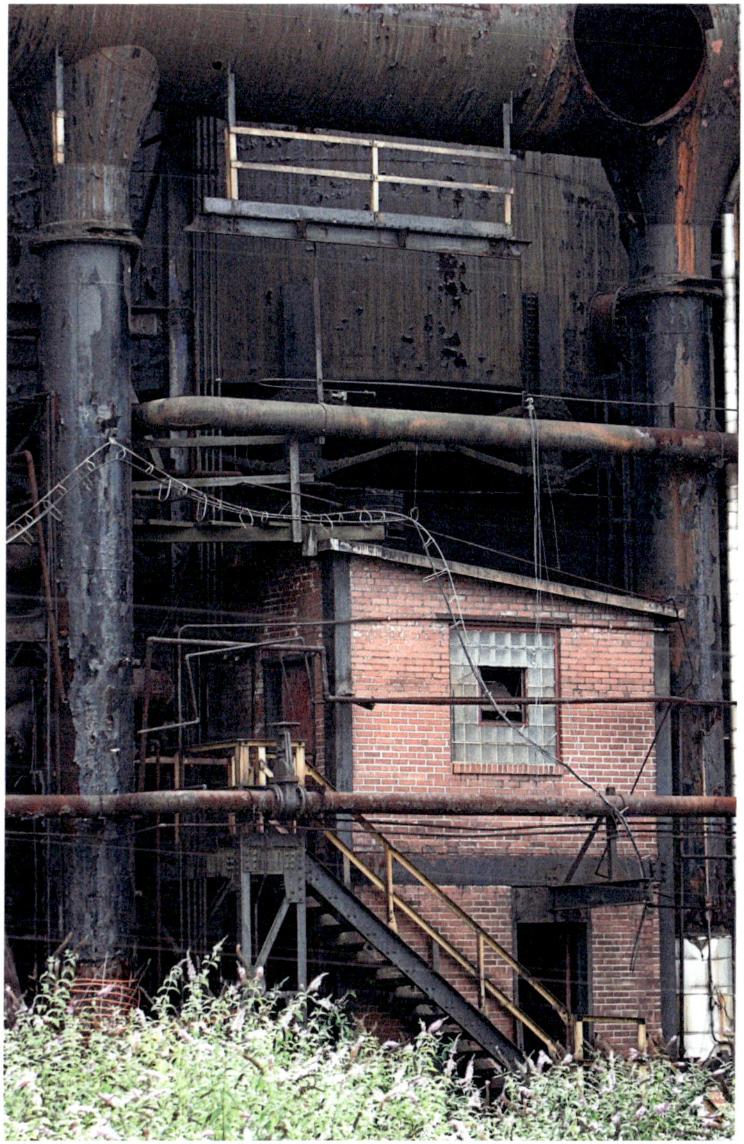

Slag Pots Underneath Blast Furnace

North end of blast furnace; slag pots underneath of cast floor; furnace cast about 6 times a day forming molten material (iron + slag); slag was poured into slag pots and iron into submarine cars and taken to the iron foundry and converted into steel.

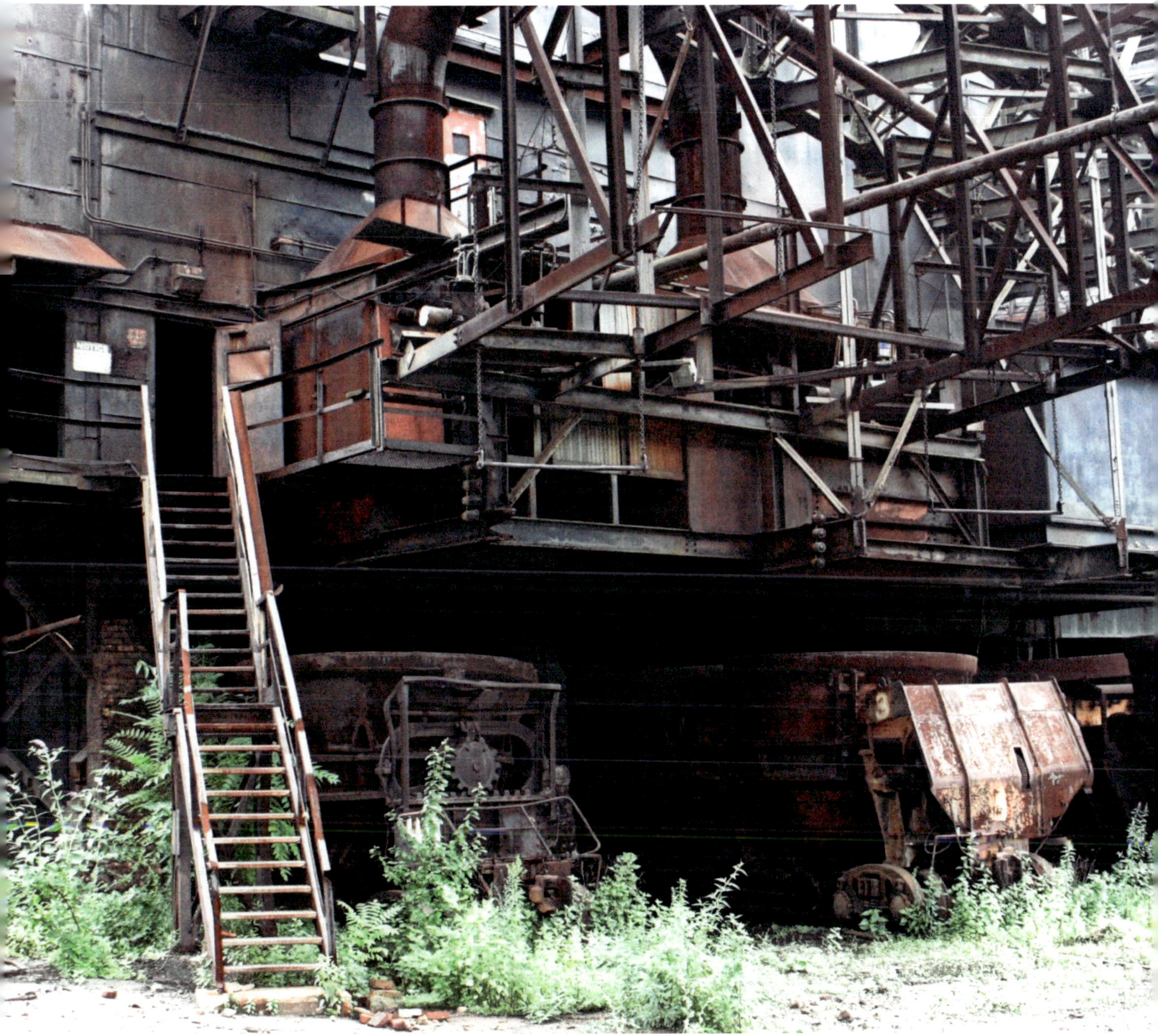

Slag Pots

Each car had two slag pots; when tipped over to pour excess slag over the bank the slag would light up the sky to a reddish orange.

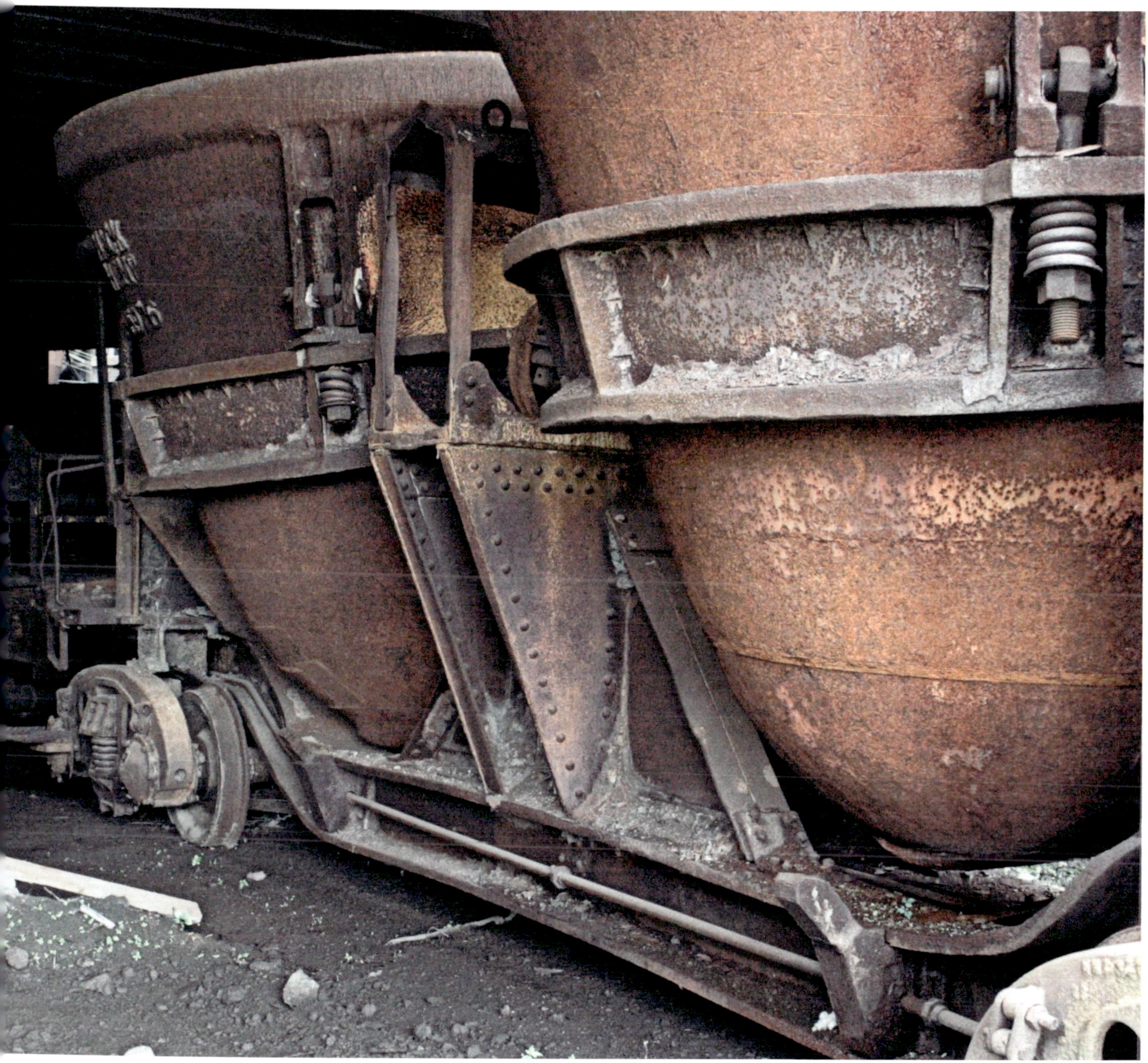

Slag Pot Detail

Showing rough detail of slag pot castings.

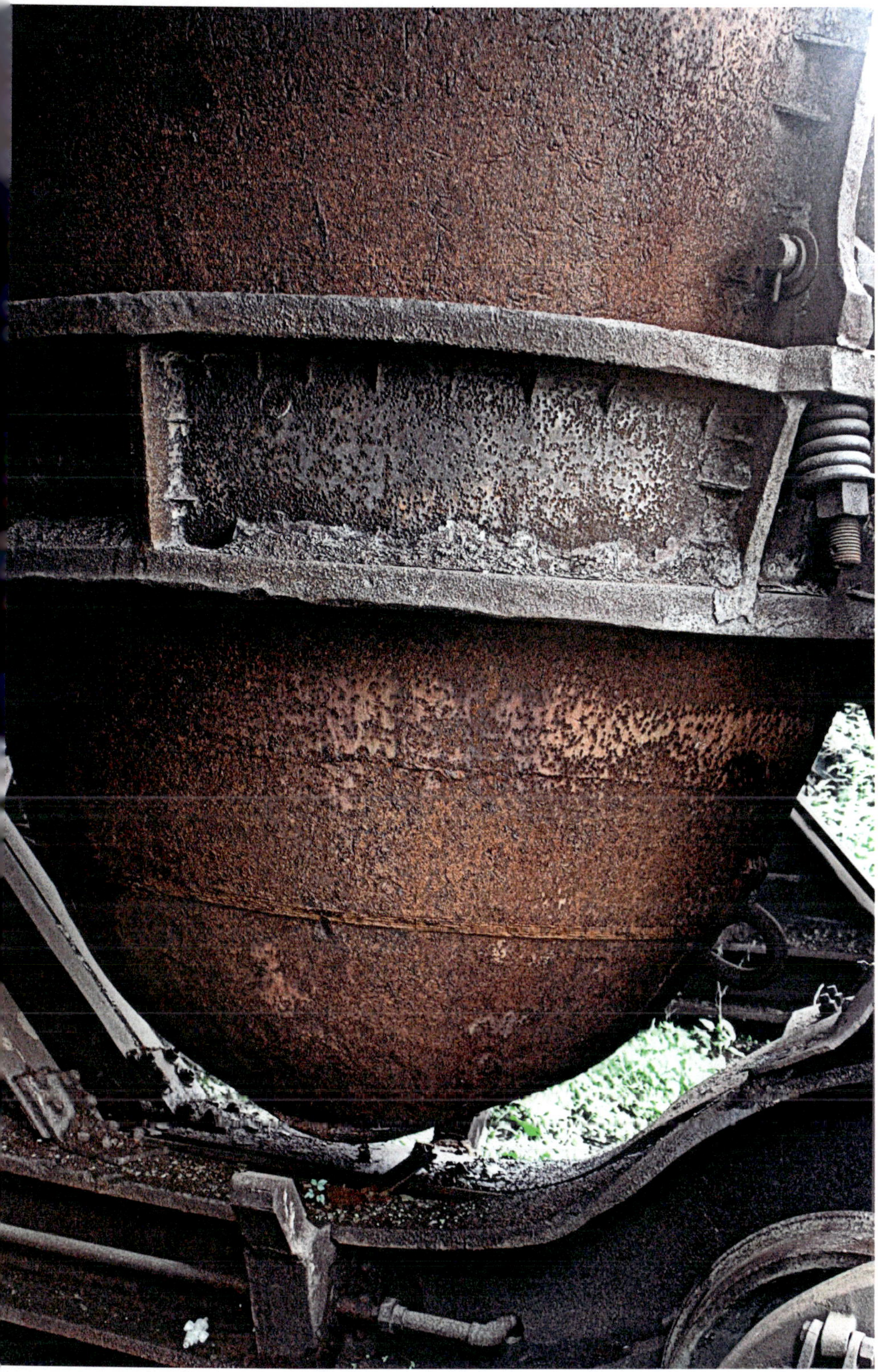

Number 2 machine shop looking west.

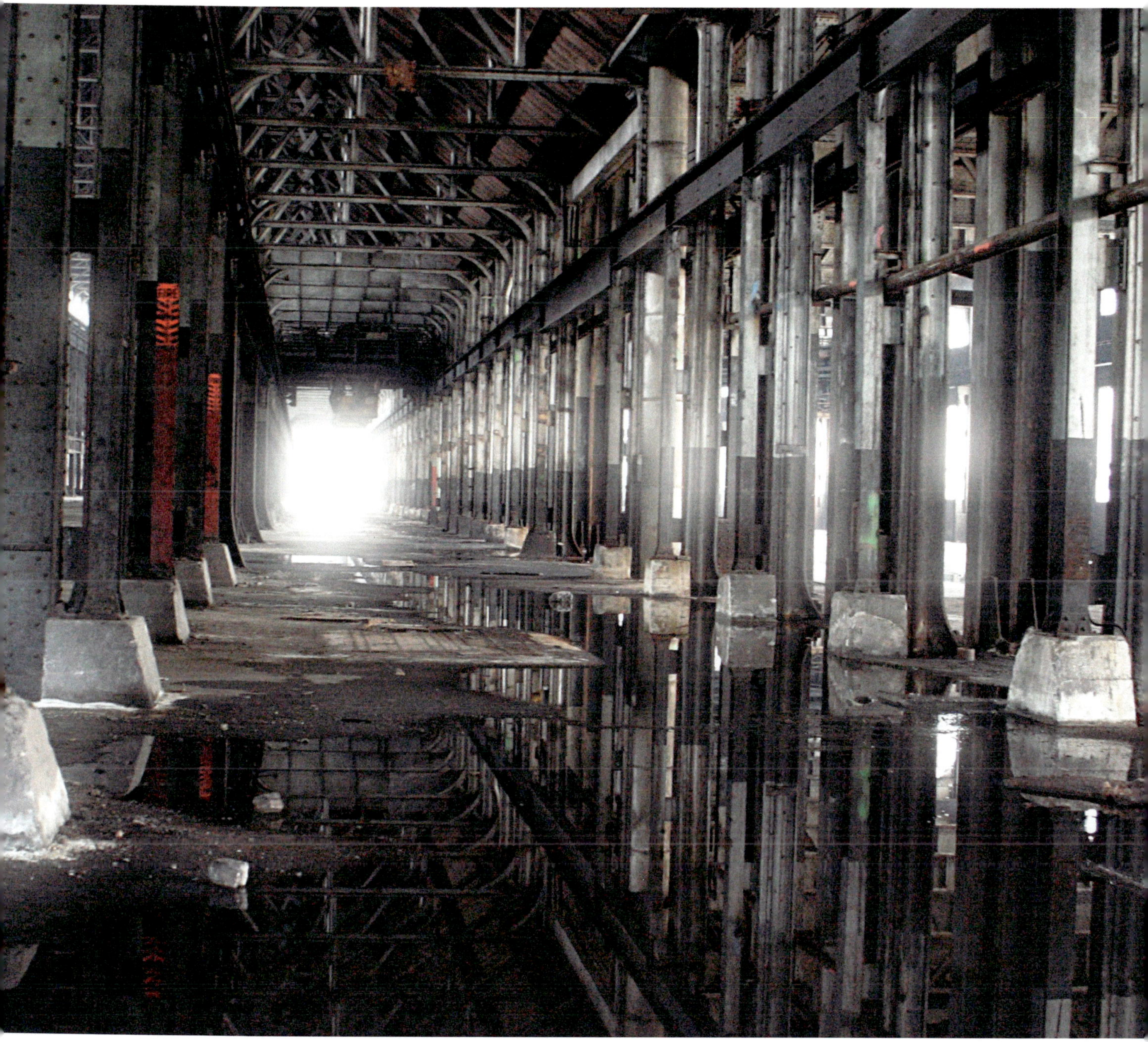

Door into Blowing Engine House.

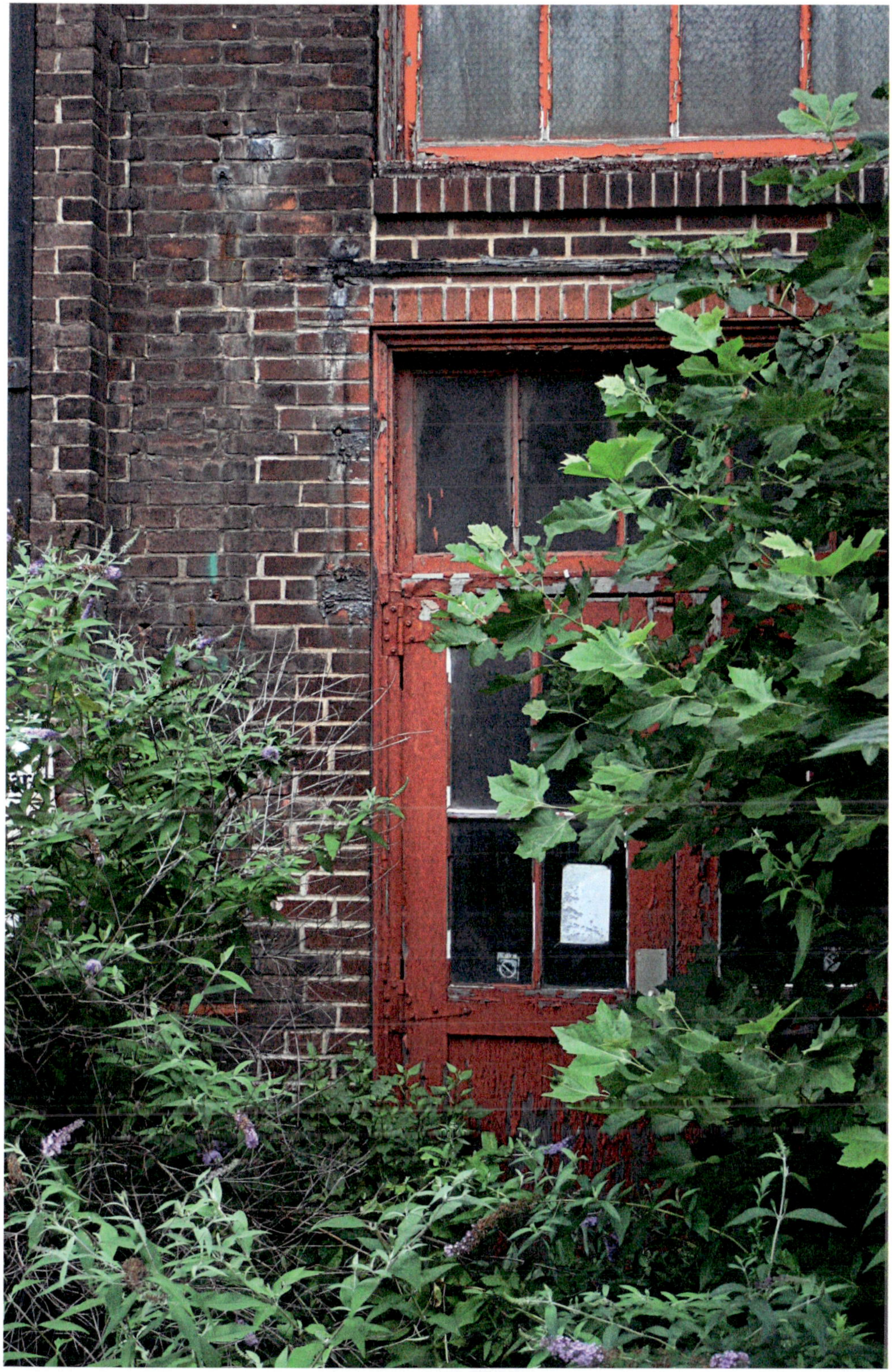

Interior windows of the Bessemer
Steel Rail producing facility.

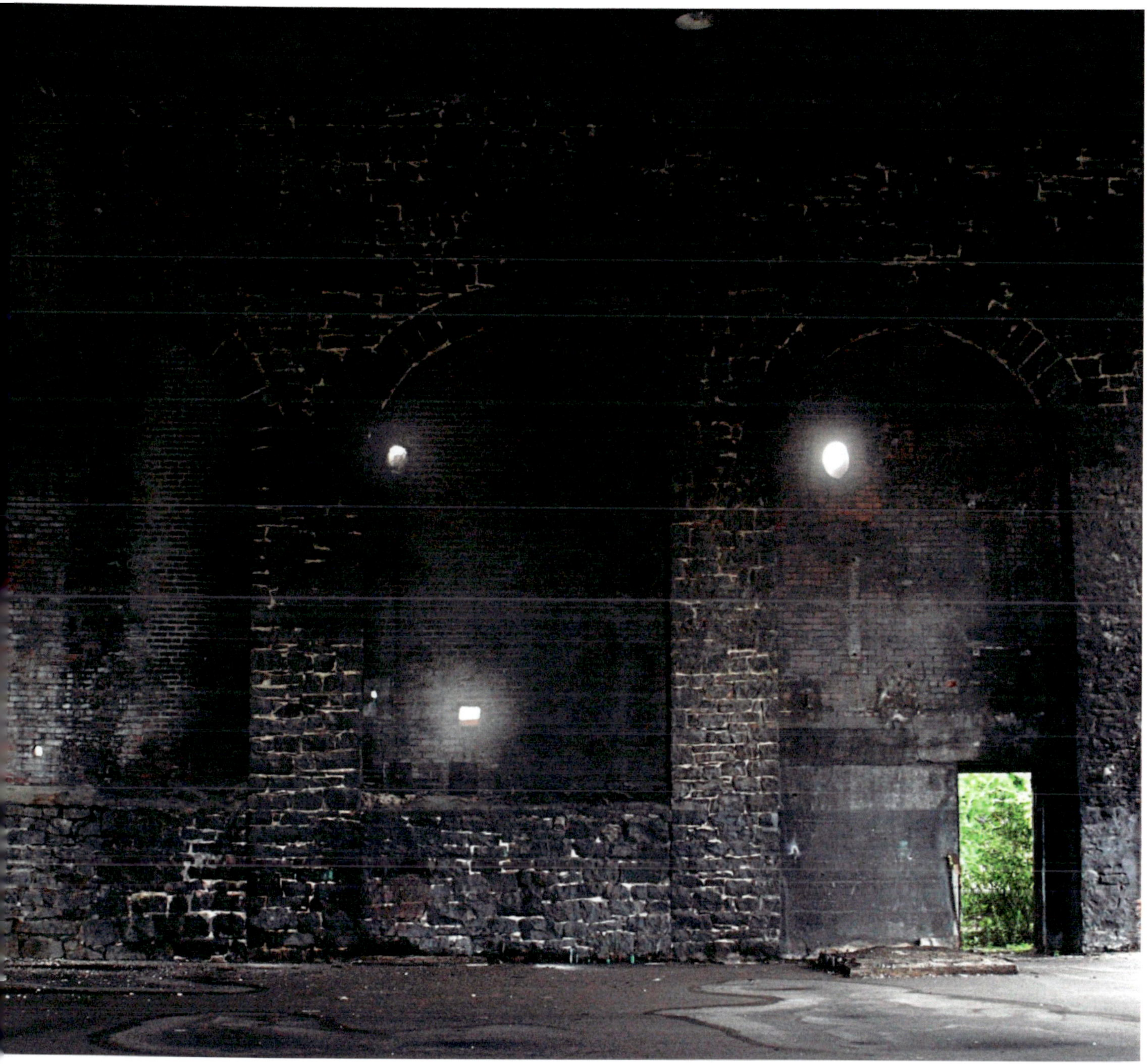

Made in United States
Troutdale, OR
11/03/2026

56844881R00059